Cashew Production, Processing and Value Addition

About the Authors

Dr. M.S.Aneesa Rani (aneesaraniprof@gmail.com) has got graduated in 1992 from Tamil Nadu Agricultural University, Coimbatore. She completed her doctoral studies in 2002 from the same campus. She has 12 years of experience in cashewnut research and development at Regional Research Station, Tamil Nadu Agricultural University, Vriddhachalam, Cuddalore district under the project ICAR-AICRP in Cashew. She has developed many varieties and technologies, among which VRI(cw)H1- the first F1 hybrid cashew in Tamil Nadu, High-Density Planting System in Cashew, Ultra-high Density Planting System in Cashew and HC 6- a dwarf cashew hybrid suitable for high density planting system are popular among the cashew farmers of Tamil Nadu. She also took strenuous efforts to popularize these varieties and technologies for the benefit of the farmers of Tamil Nadu, for which, she has been awarded with Young Women Scientist Award 2010 by Science City, Ministry of Higher Education, Government of Tamil Nadu, Best Women Scientist Award 2011 and Best Research Station Award 2016 by Tamil Nadu Agricultural University, Coimbatore. She has published 6 international, 10 national papers, 20 books and booklets to add to her crown. Presently she is working at TNAU Information and Training Centre, Annanagar, Chennai.

Dr. K.Murali, Ph.D.,(muralihort@yahoo.com) is working as Horticultural officer at Ulundurpet, Tamil Nadu. Graduated M.Sc. and Ph.D degree in TNAU, Coimbatore. Earlier he worked as JRF and SRF in TNAU. He published 2 articles at International level and 1 article at National level. He has received DAAD fellowship at Leibniz Universitat, Hannover, Germany during his study period.

Dr. D.Keisar Lourdusamy, Ph.D. (keisar@tnau.ac.in)) is working as Associate Professor of Horticulture at Horticultural College and Research Institute, TNAU, Periyakulam. Earlier he worked in All India Coordinated Research Programme on Cashew at Regional Research Station, TNAU, Vriddhachalam as Junior Horticulturist. He was one of the member of research team working on crop improvement, crop management and crop protection of cashew and organized Research seminars, District and State level seminars and Training programmes in Tamil Nadu and made several publications on cashew along with his colleagues.

Cashew Production, Processing and Value Addition

Dr. M.S. ANEESA RANI, Ph.D.

Dr. K. MURALI, Ph.D.

Dr. D.KEISAR LOURDUSAMY, Ph.D.

SSPH

SATISH SERIAL PUBLISHING HOUSE

403, Express Tower, Commercial Complex
Azadpur, Delhi-110033 (India)
Phone : 011-27672852, Fax : 91-11-27672046
e-mail : info@satishserial.com, hkjain1975@yahoo.com
Website : www.satishserial.com

Published by :

SATISH SERIAL PUBLISHING HOUSE
403, Express Tower, Commercial Complex, Azadpur, Delhi-110033 (INDIA)
Phone : 011-27672852 Fax : 91-11-27672046
E-mail : info@satishserial.com, hkjain1975@yahoo.com

ISBN 978-93-88020-27-5

Composed, Designed & Printed in India

भाकृअनुप
ICAR

ICAR - DIRECTORATE OF
CASHEW RESEARCH, PUTTUR

काअनि
DCR

Dr. M. Gangadhara Nayak
Director & PC-AICRP Cashew

Foreword

Cashew is admired as the wonder nut among all the nut crops. Cashew industry in India alone provide livelihood to one million people. Export of cashew kernels from India brings a revenue of more than Rs.5000 crores per year as foreign exchange. There are around 3800 cashew processing industries in India especially in Kerala, Karnataka and Tamil Nadu. India ranks first in cashew nut processing but third in raw nut production. New entrants in cashew cultivation like Vietnam and Nigeria are posing a threat in world market with their increased production and processing of raw cashew nuts.

Since cashew is considered as a poor man's crop, little attention was given for its cultivation. Following cultivation with modern technologies and with improved varieties, there is a possibility of doubling the raw nut production in India. The installed capacity of Indian cashew processing industry is around 20 lakh tonnes of raw nuts per year. But, the domestic production is supplying just one third of the demand of these processing industries. Hence, around ten lakh tonnes of raw nuts valued for more than six thousand crores are imported to run these processing industries.

Under All India Coordinated Research Project on Cashew, cashew research centres in different parts of India have released several high yielding varieties and technologies to improve the yield potential to more than 2000 kg per hectare. A thorough understanding of the needs of this crop, its response to biotic and abiotic factors and future research needs for sustainable yield and quality under varying climatic conditions is required for improving cashew production in the country.

The authors have taken efforts to study and analyse the various practical aspects of technologies related to cashew research to improve the production in Tamil Nadu for about 12 years. As a leading research centre for cashew in India, Vriddhachalam centre released four varieties namely VRI 1, 2, 3, 4 and one hybrid, VRI(cw)H1 in cashew for the development of farming community in Tamil Nadu.

I appreciate their efforts for having collected the research and development activities related to cashew and publish in the form of a book and I definitely feel that this book will be useful for the cashew researchers, growers and all stakeholders in India.

M. Gangadhara Nayak

Preface

India has been the prime producer and exporter of cashew in the world, earning a forex of about Rs. 5000 crores every year. Cashew industry has emerged as one of the potential enterprises in accelerating the economic growth of the country. India is the world leader in Cashew production and exports but its position is now being challenged by Vietnam, Brazil, Kenya and Tanzania. In order to meet the challenge, the rate of growth in production should be accelerated by adopting intensive development strategies by bringing more area under cashew and replanting senile plantation with clones of high yielding varieties and also intensive transfer of technology programmes.

High yielding varieties and improved technologies available for boosting cashew production are compiled in this book **'Cashew Production, Processing and Value Addition'** which will be useful for all those dealing with this crop and will serve as a useful reference material for students, farmers, extension workers and policy planners.

Scientific information generated by cashew researchers have been liberally used for which the authors gratefully acknowledge. The authors are grateful to Directorate of Cashew Research (DCR), Puttur, Project Coordinator - All India Coordinated Research Project on Cashew and Tamil Nadu Agricultural University for the research support.

Authors

Contents

About the Book

Cashew nut industry has emerged as one of the potential enterprises in accelerating the economic growth of the country. India has been the leader in cashew production and export for many centuries but its position is being challenged a short while ago by Vietnam, Brazil, Kenya and Tanzania. To meet out the challenges faced, this book "Cashew Production, Processing and Value Addition" has been compiled by the authors. The book consists of scientific information generated by cashew researchers which were generally used under topics namely cashew global and national scenario, crop improvement, crop production and management, plant protection, processing of nuts, value addition, marketing and strategies for development of cashew nut. The information compiled in the book will be useful for all the people dealing with this crop and will serve as a useful reference material for students, farmers, extension workers and policy planners to work towards improving productivity and export of cashew nuts in the country.

Chapter 1

Cashew Global and National Scenario

1.1 History and uses of cashew

Cashew (*Anacardium occidentale* L.) is an export-oriented crop grown for its nuts, a true drupe, often referred as 'wonder nut'. It is one of the most valuable processed nut traded on the global commodity markets and is also an important cash crop. It has the potential to provide source of livelihood for the cashew growers, empower rural women in the processing sector, create employment opportunities and generate foreign exchange through exports. The term 'Cashew' has originated from the Brazilian name 'acajaiba' and the Tupi name 'acaju', which the Portuguese converted into 'caju' and is commonly known as 'kaju' in India. It is known as 'Paragi Andi' in Kerala meaning foreign nut, 'Lanka Beeja' in Odisha assuming its introduction from Sri Lanka, and 'Mundiri' indicating the shape of the nut in Tamil Nadu.

Being an evergreen tree of tropics, this is cultivated in tropical region on either side of the equator for its delightful nutritious kernels and apple, besides its by-product like cashew nut shell liquid (CNSL). Its nutritional composition compares well with almond, hazel nut and walnut. Needless to mention that delicious cashew kernels are excellent dietary supplements owing to their unique combination of fat (47–50.9%), proteins (21%), carbohydrates (18-20%), minerals, vitamins and dietary fiber and thus, find place from tasty dry nuts, sweet dishes to a range of confectionary products all over the world. Cashew apples, the pseudo fruits, rich in ascorbic acid (40–390 mg g^{-1}), sugars, minerals and other antioxidants are used in limited quantities for preparing concentrated juice, syrup, ready to drink beverage, jam, nectar, frozen pulp, fruit wafers, wine, vinegar, cashew feni, *etc.* CNSL finds extensive use in paints, textile, resin industries and ship building industries and thus, is considered as an important by-product of high trade value. It is used to cure scurvy, diarrhea, uterine complaints and dropsy.

1.2 Origin and Distribution

Cashew is native to Brazil. It was introduced into India by Portuguese travellers in the 16th century for afforestation, soil conservation and waste land development. Subsequently it was dispersed to other parts of the country and also to South - East Asia. Around same time, it was introduced to East African countries (Peter, 2002). India was the first country in the world to exploit international trade in cashew kernels in the early part of 20th Century.

Cashew nut is processed in almost all the continents but cultivated only in some limited countries. The cultivation has been distributed in Asian countries such as all South Indian states in India, Vietnam, Thailand, Indonesia, West and East African countries and Brazil. In the African zone, Nigeria, Côte d'Ivoire and Tanzania are the major producers, besides other countries like Benin, Guinea Bissau, Mozambique, Ghana, Senegal and Madagascar. The primary producers in the Latin American zone is Brazil, besides Columbia, Costa Rica, Honduras and Salvador (Fig. 1). The nut is also cultivated in some packets of Sri Lanka, Australia, etc. In India, Kerala, Karnataka, Goa and Maharashtra on West Coast, and Tamil Nadu, Andhra Pradesh, Odisha and West Bengal on East coast are the major cashew growing states.

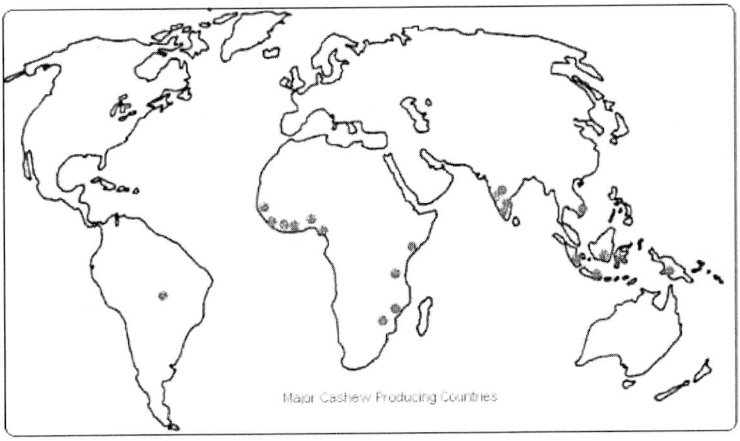

Fig. 1. Major Cashew Producing Countries

1.3 Status of Area, Production and Productivity

India is the leading country in the world in cashew area (10,27,200 ha) and production (7,25,420 MT). Cultivation of cashew in India is confined mainly to the peninsular areas. It is grown in Kerala, Karnataka, Goa and Maharashtra along the west coast of the country and in Tamil Nadu, Andhra Pradesh, Odisha and West Bengal along the east coast of the country. To a limited extent it is being cultivated in Chhattisgarh, North Eastern States (Assam, Manipur, Tripura, Meghalaya and Nagaland) and in Andaman and Nicobar Islands (Table 1,2 and Fig. 2).

Table 1. Area, production and productivity of cashew in the different states of India (2014-15).

State	Area in '000 Ha.	Production in '000 MT	Average Productivity in Kg per Hectare
Kerala	84.53	80.00	946
Karnataka	124.71	80.50	645
Goa	58.17	32.00	550
Maharashtra	186.20	235.00	1262
Tamil Nadu	140.42	67.00	478
Andhra Pradesh	185.45	100.00	539
Odisha	180.41	85.50	474
West Bengal	11.36	13.00	1096
Others	56.15	32.42	658
Total	**1027.20**	**725.42**	**706**

Source: http://dccd.gov.in/Content.aspx?mid=1075&tid=1

Fig. 2. Major cashew growing areas in India

1.3.1 Major cashew growing areas in India

Cashew cultivation in India is concentrated mostly in Western and Eastern coasts comprising of Kerala, Karnataka, Goa, Maharatstra (West Coast), Tamil Nadu, Puducherry, Andhra Pradesh, Odisha and West Bengal (East Cost). To a small extent cashew is grown in Madhya Pradesh, Manipur, Tripura, Meghalaya and Andaman Nichobar Islands.

- **Kerala :** Cannanore, Malapuram, Palghat, Trichur, Quilon, Trivandrum, Calicut, Ernakulam, Alleppy, Pathanamthitta, Kottayam, and Iddukki.
- **Karnataka :** Dakshina Kannada, Uttar Kannada, Shimoga, Kolar, Belgum, Chikmagalur and Kodagu.
- **Tamil Nadu :** Cuddalore, Ariyalur, Tiruchirapalli, Pudukottai, Sengalpattu, Thanjavur, Theni and Kanyakumari.
- **Puducherry :** Puducherry and Karaikkal.
- **Andhra :** West Godavari, East Godavari, Srikakulam, Visakhapatnam, Vijayanagaram, Krishna and Prakasam.
- **Maharashtra :** Ratnagiri, Sindhudurg, Kolhapur and Kolaba.
- **West Bengal :** Midnapore (East), Midnapore (West), Bankura and Burdwan.
- **Odisha :** Dhenkanal, Koraput, Cuttack, Puri, Ganjam, Sambalpur, Balasore and Sundargarh.

1.4 Export scenario

India has a creditable record of attaining good foreign exchange by exporting cashew kernels. During the year 2014-2015, India exported 1,18,952 MT of cashew kernels valued at Rs. 5432.85 crores (Table 2). USA, The Netherlands, UK, Japan, UAE, France, Canada, Saudi Arabia, Singapore, Italy, Germany, Austria, Israel and Spain are the major international buyers of Indian cashews. Further, cashew nut is also imported into India from other producer countries. These nuts are processed in the country and then are either distributed in the local market or re-exported.

Table 2. Export of Cashew Kernel, Cashewnut Shell Liquid (CNSL) & Import of Raw Cashewnut

Year	Cashew Kernel Export		CNSL Export		Raw nut Import	
	Quantity (MT)	Value (Rs. Lakhs)	Quantity (MT)	Value (Rs. Lakhs)	Quantity (MT)	Value (Rs. Lakhs)
2000-2001	89155	2,04,975	2246	389	2,49,318	96080
2001-2002	97550	1,77,680	1814	419	3,56,566	96001
2002-2003	104137	1,93,302	7215	925	4,00,659	1,23,657
2003-2004	100828	1,80,442	6926	703	4,52,898	1,40,093
2004-2005	126667	2,70,924	7474	791	5,78,884	2,18,324
2005-2006	114143	251486	6405	709	565400	216295
2006-2007	118540	245515	5589	920	592604	181162
2007-2008	114340	228890	7813	1197	605970	174680
2008-2009	109522	298840	9099	2606	605850	263241
2009-2010	108120	290582	9748	2412	752894	303735
2010-2011	105755	281939	12051	3377	529370	264956
2011-2012	131760	439068	13575	5946	809371	533776
2012-2013	104015	406721	9192	2984	892160	533112
2013-2014	114791	505873	9480	3861	771356	456399
2014-2015	118952	543285	10938	5581	939912	657093

http://dccd.gov.in/Content.aspx?mid=1077&tid=1

The first commercial cashew processing unit was set up in Kollam in the mid-1920s. However, early exports were not followed up because the cashew kernels were not vacuum-packed, but wrapped in newspapers and packed in reused tea chests. World War II put a halt to any further development of the trade. Export volumes picked up only after the introduction of airtight tins infused with carbon dioxide in the mid-1950s. Soon after this technological advancement, processing plants were established in Mangalore and Goa. Another small but important processing centre is located in Panruti in Cuddalore district of Tamil Nadu. A major problem facing the Indian cashew industry is the acute shortage of raw cashew nuts within the country. The processing requirement is about 1,643,000 MT per annum while the domestic production is only 725,000 MT per annum. Consequently, the cashew processing industry has resorted to importing raw nuts for processing and export. Along with the increasing world demand for cashew, Indian consumption is also growing (Table 2).

India and Brazil were the major suppliers of cashew to the world market until about five years ago. In recent years, Vietnam has also emerged as a major supplier, replacing Brazil in second position. At present, USA and West Europe are India's major markets followed by Japan, West Asia and Australia. New emerging markets like Eastern Europe, Commonwealth of independent state countries and China are becoming active buyers of cashews. Another area where there is potential for significant market growth is export of organically produced cashews. Moreover, new products are being developed apart from the cashew kernels, including cashew apple, cashew nut shell liquid (CNSL), cashew shell and cashew testa.

Table 3. Cashew processing units in India

States	Processing Units (Nos.)	Processing Capacity	Utilization ('000 MT) Indigenous	Import	Total
Kerala	432	600	67	320	387
Karnataka	266	300	45	20	65
Goa	45	50	21	–	21
Maharashtra*	2200	50	20	–	20
Tamil Nadu	417	400	294	225	519
Andhra Pradesh	175	100	92	–	92
Odisha	209	100	11	–	11
West Bengal	30	8	8	–	8
Chattisgarh	3	5	–	–	–
NE States	22	10	15	–	15
Total	3799	1623	573	565	1138

* Includes 1850 small scale cottage industries

Cashew apples can be sold fresh as soon as they have been picked and are then used as a culinary ingredient or further processed into drinks (juice, wine), marmalade or vinegar. There are interesting prospects for setting up processing plants for utilization of cashew apples, which at present, are mostly wasted. Many preparations like juices, jams, candies, pickles, chutneys and alcoholic beverages can be prepared from this by-product.

Cashew nuts consist of 35-45% seed and around 55-65% shell. The shells contain 15-30% oil. A ton of nuts contains around 200 kg kernel and 180 kg oil (cashew nut oil or CNSL). CNSL is a naturally occurring phenol, which is contained in the soft honeycomb mesocarp of the shell. CNSL is a viscous, oily or balsam-like substance, pale yellow to dark brown, having bitter taste and caustic properties. Fresh CNSL contains 90% anacardic acid, which is converted into cardanol on heating. The remaining 10% is cardol, which is mainly responsible for the vesicant property. CNSL is a by-product of cashew processing, is used as oil in industry. It is a versatile industrial raw material that has applications in polymer-based industries such as friction dust, brake linings, paints and varnishes, laminating resins, cashew cements, polyurethane based polymers, surfactants, and epoxy resins (Peter, 2002). Cashew shell, after removal of the kernel and extraction of shell liquid, is currently used as fuel. However, this shell can also be used to manufacture particle-based boards for the packaging industry. Tannin can also be extracted from the testa of the kernel. It has application in the leather industry and many other chemical industries. The major products for industrial applications are cashew lacquer, insulating varnishes, electrical windings and electrical conductors impregnated with CNSL and cashew cement (CNSL reacted with formaldehyde).

1.5 Cashew Research in India

Cashew research in India started way back in 1950s through launch of *ad-hoc* schemes sanctioned by Indian Council of Agricultural Research (ICAR). Cashew research got further impetus through formation of Central Plantation Crops Research Institute and, later, establishment of an independent National Research Centre for Cashew (NRCC), Puttur, in Karnataka in 1986. Location-specific research programmes in the eight cashew growing states are conducted through All India Coordinated Research Project on Cashew (AICRP on Cashew) whose Headquarters are also located at the Directorate of Cashew Research (DCR) (formerly National Research Centre for Cashew), Puttur.

There were eight centres and one sub-centre under AICRP on Cashew, located all over the country till the end of 10th Five Year Plan. In the year 2009 during 11th Plan, one centre each in Gujarat and Jharkand and three co-operating centres (Arabhavi in Karnataka, Goa and Tura - Barapani in Meghalaya) were included under AICRP on Cashew. With combined efforts of the Directorate of Cashew Research, Centres of the AICRP-Cashew and SAUs, over 40 high yielding cashew varieties have been developed and released in the country.

1.5.1 AICRP on cashew Centres

1. Cashew Research Station, (Dr. YSRHU), Bapatla-522 101, Andhra Pradesh
2. Cashew Research Station, (OUAT), Bhubaneswar-751 003, Odisha
3. Agricultural Research Station, (UAS), Chintamani-563 125, Kolar District, Karnataka.
4. SG College of Agricultural and Research Station, (IGAU), Chhattisgarh
5. Regional Research Station, (BCKV), Jhargram-721 507, West Bengal
6. Cashew Research Station, (KAU),Madakkathara-680 651, Kerala
7. Regional Agricultural Research Station, (KAU), Pilicode-671 353, Kerala.
8. Regional Fruit Research Station, (Dr. BSKKV), Vengurla-416 516, Maharashtra.
9. Regional Research Station, (TNAU), Vridhachalam-606 001, Tamil Nadu.
10. Zonal Research Station, (BAU), Darisai, East Singhbhum Dist., Jharkhand.
11. Agricultural Experimental Station (NAU), Paria-396 145, Valsad District, Gujarat.

1.5.2 Co-operating Centres

1. Kittur Rani Chennamma College of Horticulture (UHS), Belgaum district, Karnataka.
2. ICAR Research Complex for Goa, Ela, Old Goa, Goa - 403 402.
3. ICAR Research Complex for North Eastern Hilly Regions, Tura-794 005, West Garo Hills District.

1.6 Institutes in cashew industry

1.6.1 Directorate of Cashew Research (DCR)

In India, research on cashew was first carried out in the early 1950s. In 1971, Indian Council of Agricultural Research (ICAR) sanctioned the All India Coordinated Spices and Cashew Improvement Project (AICS and CIP) with its headquarters located at Central Plantation Crops Research Institute (CPCRI), Kasaragod. The National

Research Centre for Cashew was established at Puttur on 18th June, 1986, after which it was upgraded and renamed by ICAR in 2009 as Directorate of Cashew Research (DCR). Subsequent to the bifurcation of AICS and CIP, the headquarters of All India Coordinated Research Project on Cashew was shifted to DCR, Puttur. At present, this Coordinated Research Project is operating in ten centers, one sub-center and three co-operating centers located in major cashew growing areas of the country.

1.6.2 Directorate of Cashew Nut and Cocoa Development (DCCD)

The Directorate of Cashew Nut and Cocoa Development (DCCD) established in 1966 under the Union Ministry of Agriculture gave a greater impetus to the development of cashew in a more scientifically oriented manner. This marked the first step towards the integration and co ordination of cashew development in association with developmental agencies of States and Research Institutes. The prime objective of DCCD is the formulation and execution of various development programmes on cashew nut and cocoa in the country. This directorate is at Kochi, Kerala.

1.6.3 Cashew Export Promotion Council of India (CEPC)

The Cashew Export Promotion Council of India (CEPC) was established by the Government of India in the year 1955, with the active co-operation of the cashew industry, to promote exports of cashew kernels and CNSL from India. Through its very set up, the Council provides the necessary institutional frame-work for performing the different functions that serve to intensify and promote export of cashew kernels and CNSL. The Council provides the necessary liaison between foreign importers and member exporters of cashew kernels. The enquiries received from foreign importers are circulated amongst Council members. The Council also extends its role to settling complaints amicably in the matter of exports/imports, either on account of quality and/or variation in fulfillment of contractual obligations. The Cashew Export Promotion Council of India is at Cashew Bhavan, Mundakkal West, Kollam-691001, Kerala.

❑❑❑

Chapter 2

Crop Improvement

2.1 Botany of Cashew

2.1.1 Taxonomy and Cytology

Cashew tree (*Anacardium occidentale* L.) belongs to family Anacardiaceae which comprises about 60 genera and 400 species of trees and shrubs with resinous bark. Though Anacardium is described as a small genus with only 8 species, over 20 species are known to exist in central and south America. The species of Anacardium vary largely with respect to size, shape, colour of peduncle and size and shape of nut and leaves. The *A. gigantium* from surinam has the biggest apple and is locally important timber in South America. The spectacular white leaves associated with the inflorescence of *A. spruceanum* make it a tree with excellent ornamental potential *A. rhinocarpus* and *A. spruceanum* possessing hard wood are useful as rootstocks. *A. excelsum* is used for construction and as a shade tree for coffee and cocoa plantations. *A. humile* a subshrub closely related to *A. occidentale* possesses edible hypocarps and seeds *A. occidentale* L. is a diploid and exhibits chromosome polymorphism. Chromosome number reported in literature ranges from 2n=24 or 42.

2.2 Morphology

2.2.1 Root system

The cashew tree has an extensive lateral root system and a taproot, which penetrates deeply into soil. The root system of the young dwarf cashew is one with very well developed main root that branches many times and can grow to 10 m or more in deep sandy soils. Lateral roots develop in the upper soil layers between 15 and 32 cm deep. The length of the superficial roots may reach twice the diameter of the crown in dry-land conditions (Barros, 1995). When irrigated the lateral roots are concentrated around the wet area of soil. The characteristics of the tap and lateral roots are of importance in relation to the fertilization of cashew. Falade (1984) when studying the effects of topography, soil texture, stoniness and the presence of a hardened soil layer on the development of the cashew root system reported great variation in the depth of the main root and distribution in depth and length of the lateral roots.

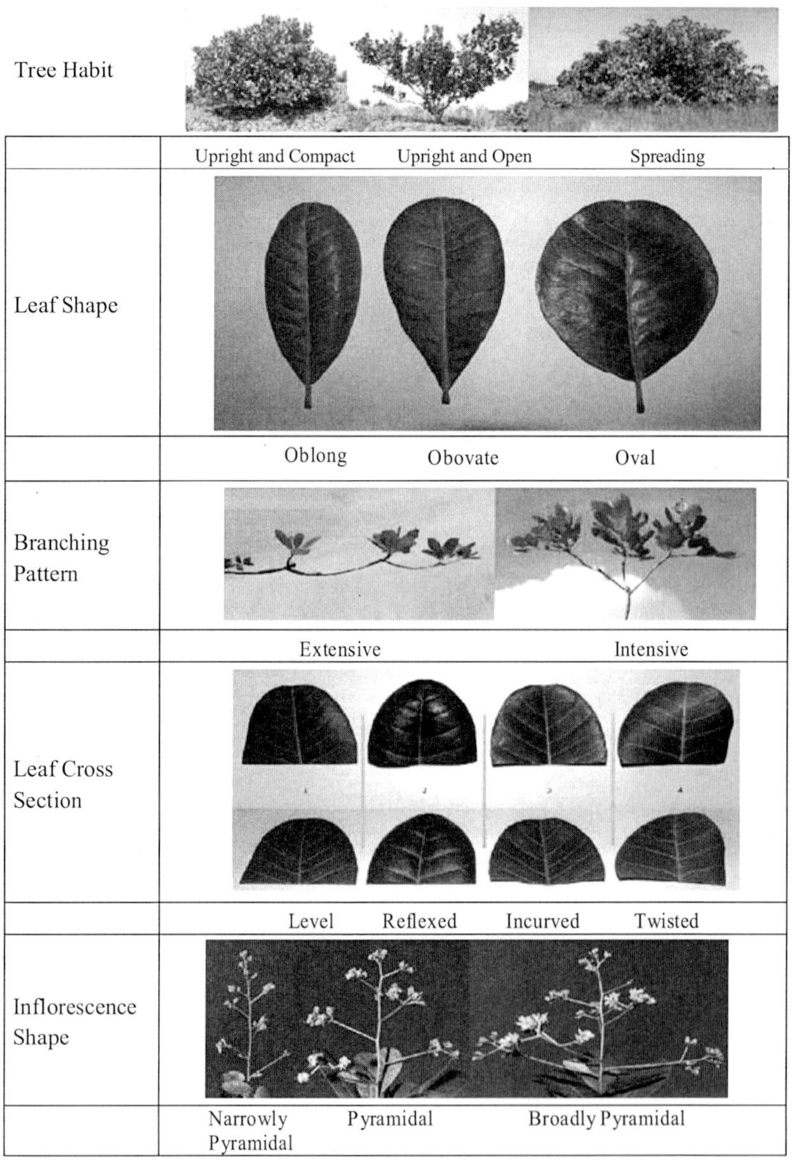

Fig. 3. Cashew Tree, branching, leaf and inflorescence (Picture courtesy: DCR, Puttur)

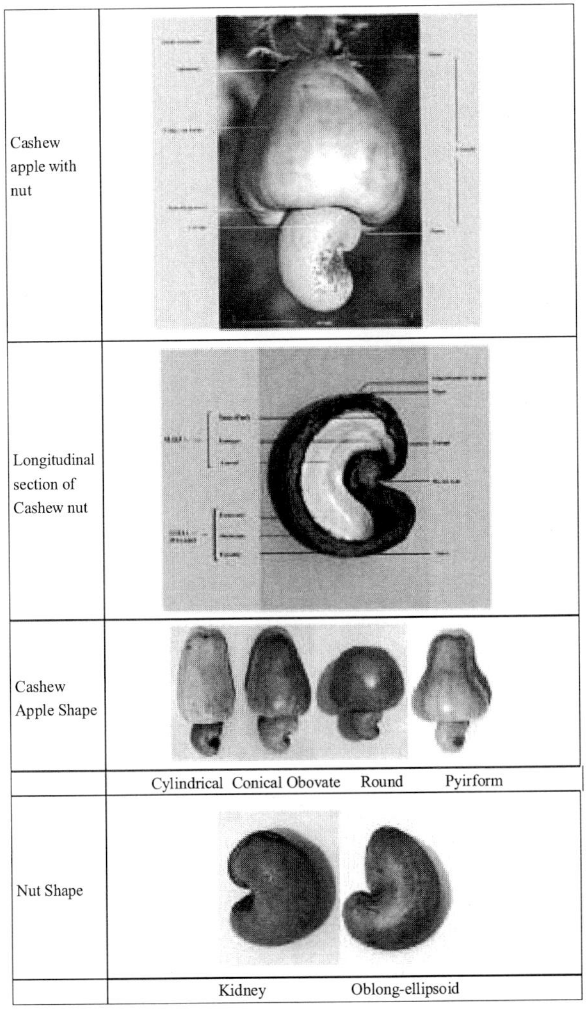

Fig. 4. Cashew apple and nut (Picture courtesy: DCR, Puttur)

2.2.2 Habit

Cashew is a spreading evergreen tree reaching upto 12 m in height. In Brazil, there are two groups of cashew. The most prevalent is the common type (giant), which grows to heights ranging from 5 to 8 m, but able to reach 15 m. The diameter of the crown generally varies from 12 to 14 m and in exceptional cases is upto 20 m (Barros, 1995). The other group, the dwarf cashew grow upto 4 m, on average, with a crown diameter of 6 to 8 m. This group of dwarf cashew flowers between 6 and 18 months (Barros *et al.*, 1998).

2.2.3 Stem

When grown under favourable conditions and unharmed by pests, its stem is erect, canopy conical or umbrella shaped. Under less favourable conditions, its tree is much smaller, stem tortuous, and branches rest on ground, creeping over soil and rooting takes place where they touch soil. This habit makes tree a very good soil conversant.

Based on the nature of branching in cashew, they are described either as 'intensive' or 'extensive'. The intensive shoot grows to a length of about 25 to 30 cm and terminates in a panicle. Simultaneously, three to eight laterals arise within 10 to 15 cm of the apex. Some of these laterals shoots may also terminate in panicles in the same flowering season, repeating the same growth pattern, giving a well covered bushy appearance to them. In the extensive type, the shoot grows to a length of 20 to 30 cm and rests. A bud sprouting 5 to 8 cm below the apex leads to further growth. This process of growth continues for two or three years without any flowering. This kind of growth pattern produces a spreading tree. Even though both kinds of branching are observed in all trees, one type dominates in a tree. High yielding trees have more than 60 per cent intensive branches, whereas low yielders have less than 20 per cent.

2.2.4 Leaves

Alternate, simple, glabrous, leathery, oblong to obovate and rounded at tip, veins prominent, 10-20 pairs, petiole swollen at base and flattened on upper surface. Young leaves are reddish-brown in certain varieties which later turn to dark green after 20 days.

2.3 Floral biology

2.3.1 Inflorescence

Inflorescence is a terminal panicle either conical or pyramidal in shape. The time from visual emergence of inflorescence to opening of first flower takes about 5-6 weeks. An inflorescence contains on an average 120-1100 flowers.

2.3.2 Flowers

Flowers are scented, white in colour at first but soon turn pink after a few days. Flowers have pubescent bract. They are andromonoecious with male and hermaphrodite flowers in same inflorescence. Each flower has 5 sepals and petals each. In male flower, there are 10 stamens of which 9 are short and one long stamen with red anther projecting above corolla. Ovary is rudimentary. In hermaphrodite flowers, 9 stamens are short as in male flower and long stamen projects just above corolla but remains below stigma. Ovary is unilocular with single ovule; style simple and exserted. The proportion of the perfect flowers varies from as low as 0.45 to 24.9 per cent in different trees. On the same tree, the perfect flowers are larger in size than the staminate. Flowering starts from December and extends upto April.

2.3.3 Pollination

Peak anthesis is between 9 am to 11 am, stigma is receptive as soon as flowers open and remains receptive for 48 hours from anthesis. Anther dehiscence takes place 1-5 hours after anthesis. The effect of plant age on pollen abortion of *Anacardium occidentale* L. was studied. Using Alexander's strain, which differentiate aborted and non-aborted pollen grains, the proportion of aborted pollen and non-aborted pollen grains, the proportion of aborted pollen versus plant age, size variation in aborted and non-aborted pollen grains per flower were estimated. Aborted pollen was smaller than non-aborted pollen regardless of plant age. The proportion of aborted pollen varied from 22.5 to 46.8% showing a steady increase with plant age and the number of pollen grains per flower also increased with age. These results support the qualitative prediction that pollen abortion gradually increases with ageing, reflecting an increase in genetic load with age (Bhattacharya, 2005).

It is a predominantly cross pollinated tree with entomophilous pollination (Nambiar and Pillai 1985). Pollination takes place through bees, which transfer sticky pollen to stigma. The flower visitors, pollination and fruit set of cashew were studied. The yield of fruits was low due to the lack of appropriate pollinators and the possible role of ants in pollen damage. Bees (*Aphis cerana indica, A. florae, A. dorsata and Bombus spp.*), flies, butterflies, beetles and ants visited the flowers. The relative abundance of visitors coincided with nectar availability. Breeding manipulation by bagging experiments indicated that bees are efficient pollinators, increasing fruit set, while ants decreased fruit set by damaging the viable pollen. Effective fruit production requires more activities of bees and management to reduce ant damages in cashew (Bhattacharya, 2004).

Hand pollination resulted in overall increase of 20.9% fruit set against 4 to 6% in natural pollination (Jayalekshmy and John, 2005). Evaluation of pollen parents as pollinate varieties revealed that BT1 is a potential pollinate type which gave 14.93% and 13.5% fruit retention in V_2 and V_5 respectively (Deshpande *et al.*, 2003).

2.3.4 Fruit set

Cashew produces 10% of perfect flowers of which 85% are fertilized and only 4 -6% is carried to final maturity, the rest being shed at various stages. It takes about 60 days from fruit set to maturity.

2.3.5 Fruit

Kidney- shaped nut called achene, greyish- brown in colour with a hard pericarp consisting of a leathery exocarp, spongy mesocarp containing CNSL and a hard and brittle endocarp. Cashew apple is a false fruit, is the swollen pedicel. It is a fleshy and juicy peduncle of varying color and size. Currently less than 20% of the total peduncles produced are used, mainly to consume fresh or made into sweets, jams and diverse drinks. Generally, the ratio of nut to peduncle is 1:10 (w/w).

2.3.6 Kernel

Kernel is kidney shaped with reddish- brown testa, two large, white cotyledons and a small embryo. The cashew nut is rich in vitamins, non-saturated fatty acids and proteins.

2.4 Conventional Breeding

Yield in cashew is a complex character involving an integrated set of attributes namely, inflorescences per unit area, nuts per inflorescence, and mean nut weight. These variables either directly or through their interaction influence total nut yield in cashew. Any attempt to improve the yield should be preceded by understanding process governing these components. The process of differentiation of reproductive shoots from vegetative shoots is an important aspect that needs to be investigated. Although data available at present are not adequate to offer an explanation of differentiation of vegetative shoots into reproductive shoots, indications are that this could be governed by environmental variables such as nutritional factors, availability of moisture and weather parameters. Studies on compatibility relationship of cashew varieties and designing models for establishing orchards with polyclones to ensure the highest compatibility and for realizing higher yields in cashew is a priority.

Tree nuts are generally considered as highly nutritive and have been placed in the base of Mediterranean Diet Pyramid developed by World Health Organization, which recommends their daily consumption. Varietal differences with respect to neutral lipids and glycolipids were reported, whereas composition of phospholipids did not differ among varieties. Some of the varieties having >35% protein, lysine >50 µg/ mg protein and >14% of sugar were identified. These can be used in the breeding program to develop varieties with better nutritive value for diet-conscious consumer markets. One of the major production constraints in India as well as in other cashew growing countries is incidence of tea mosquito bug (TMB) in flushing and flowering season. Studies made so far on screening for resistance indicated that among germplasm accessions available in the country. Keeping this in view, the following breeding objectives evolved in different cashew research station and several varieties are released so far.

2.4.1 Breeding objectives

- Dwarf and compact canopy
- Intensive branching type
- Short flowering and fruiting phase
- More than 20 per cent perfect flowers
- 8-10 fruits per panicle
- Medium to bold nuts
- Shelling per cent > 28%
- High yield–20 kg/tree/year
- Biotic and abiotic resistance.

2.4.2 Germplasm evaluation and selection

Early attempts for germplasm collection in India were made during 1952- 1957 with sanctioning of ad-hoc schemes in Kerala, Karnataka, Andhra Pradesh, Assam and Maharastra. A total of 1490 germplasm accessions are conserved at National Research centre on Cashew at

Puttur and a different cashew research stations in India. These are primarily indigenous types selected from seedling progenies of limited initial introductions with a few exotic types from Brazil, Nairobi, Lindi, Nacala, Mozambique, Singapore, Australia and Republic of Panama. Germplasm collections also include allied species of *Anacardium* such as *A. microcarpum, A. pumilum and A. orthonianum.*

In-situ conservation of cashew germplasm is done only in amazon forests of Brazil, the original home of cashew. Subsequent to establishment of NRCC at puttur in 1986, germplasm collection through seeds was discontinued. In the National gene Bank of NRCC, Puttur, *ex-situ* conservation of 433 clonal germplasm collections is done. Similarly, Regional Cashew Gene Banks is established at Vengurla, Vridhachalam, Bhubaneswar, Madakathara and Cintamani. So far 255 accessions are characterized and catalogued as per International Plant Genetic Resources Institutes (IPGRI) cashew descriptors and the catalogue of minimum Descriptors of cashew germplasm accessions-I, II, III is published (Rao and Swamy, 2006). the catalogue of minimum Descriptors of cashew germplasm accessions–IV and V were also recently added to this list. (Nayak *et al.,* 2014).

2.4.3 Germplasm holding in India

Table 4. Cashew germplasm maintained in different states

State	Research station	No. of accessions	
		Seedling	Clonal
Andhra Pradesh	Cashew Research Station , Bapatla	87	26
Karnataka	CPCRI, Vittal	292	-
	NRCS, Puttur	-	300
	ARS, Ullal	111	-
	ARS, Chintamani	75	-
Kerala	CRS, Anakkayam	84	-
	CRS, Madakathara	93	-
Maharashtra	Regional Fruit Research station, Vengurula	80	77
Odisha	CRS, Bhubaneswar	67	-
Tamil Nadu	RRS, Virudhachalam	219	68
West Bengal	RRS, Jhargram	81	8
	Total	1189	547

In recent years, with increasing concern for quality, emphasis has been placed on identification of varieties with kernel weights of over 2g falling into export grade of W 210 to W 240. To realize higher recovery of whole kernels, standards are fixed for shelling percentage to be not less than 30%. This calls for identification of donor parents having these characters that can transmit them to their progeny. Immediate priority of Indian cashew germplasm programme is to enhance genetic variability through introduction of exotic types from Central America and Brazil, where diverse types including dwarf ones exist.

At the Regional Research Station, Vriddhachalam cashew germplasm collection has been made since 1963. The accessions were collected from the forest plantations, farmers' fields, and research stations of Tamil Nadu, Kerala, Karnataka, Andhra Pradesh and few from other countries. In the Regional Cashew Gene Bank (RCGB), the clonal accessions were panted during 1994. Cataloguing was done in 125 cashew germplasm accessions. These accessions have been grown in rain fed condition by adopting the recommended package of practices. The cataloguing has been done in accordance with the IBPGR descriptors. The catalogue contains characterization data expressed in 68 data fields (01-68). 106 accessions have upright and compact tree habit (84.8%) and the branching habit was extensive in 98 accessions. Maximum number of accessions had mid-season flowering and 78 accessions had compact inflorescence and medium sex ratio. The nut weight was intermediate in 59 accessions (47.2%) and low in 64 accessions. 21 accessions had medium cumulative nut yield and 104 accessions had low cumulative nut yield. (Jeeva *et al.,* 2015).

2.4.4 Variability recorded in seedling progenies

Table 5. Variability and its range

S.No.	Character	Range
1.	Flowering season	October to January
2.	Flowering duration	40 -127 days
3.	Harvesting duration	30 -105 days
4.	No. of fruits/panicle	1-8
5.	Apple weight	30 -150 g
6.	Nut weight	2.4 -18 g
7.	Apple : Nut ratio	4:1-12:1

8.	Shelling percentage	19 -35
9.	Kernel weight	0.5 to 4.5
10.	Shell thickness	1.5 - 5.0mm
11.	Mean yield/plant/year (10 years mean)	0.5 - 11.75 kg

Source: Rao and Swamy, 2006.

Since cashew is primarily a cross pollinated crop, it is highly heterozygous and segregation in considerable variation in seedling population. An ideal cashew plant should have dwarf and compact canopy with intensive branching habit, short flowering and fruiting phase >20% perfect flowers, 8-10 nuts/Panicle, medium to bold nuts (8-10g) with shelling percentage of >28, high yield potential (>20 kg/ tree/year) and tolerance to major pests and diseases. Evaluation of seedling progeny at different cashew research stations resulted in identification of superior genotypes for several economic characters. So far, 25 high yielding varieties were released through selection (Table 6).

Table 6. Cashew varieties developed through selection in India

Variety	Source of germplasm	Institution	Yield kg /plant	Nut weight (g)	Kernel weight (g)	Shelling (%)	Grade
NRCS-1	VTH 107/31	NRCS, Puttur	10	7.6	2.2	28.8	W210
NRCS-2	Simha-chalam	NRCS, Puttur	9.0	9.2	2.5	28.6	W210
Ullal-1	Thalipar-amba 8/46	ARS,Ullal, Karnataka	16	6.7	2.1	30.7	W210
Ullal-2	3/67 Guntur	ARS,Ullal, Karnataka	9.0	6.0	1.8	30.5	W240
Ullal-3	5/37 Manjeri	ARS,Ullal, Karnataka	14.7	7.0	2.0	30.7	W210
Ullal-4	2/77 Tuni	ARS,Ullal, Karnataka	9.5	7.2	2.2	31.0	W210
Ullal-5	2/27 Nileshwar	ARS,Ullal, Karnataka	10.5	9.0	2.9	32.8	W180
Chinth-amani	18/46 Taliparamba	ARS, Chinthamani	7.2	6.9	2.1	31.0	W210
BPP3	3/3 Simha-chalam	CRS, Bapatla	11.0	4.8	1.3	28.1	W400

BPP4	9/8 Epur-upalem	CRS, Bapatla	10.5	5.2	1.2	23.0	W400
BPP5	Tree 1	CRS, Bapatla	11.0	5.2	1.2	24.0	W400
BPP6	Tree 56	CRS, Bapatla	10.5	5.0	1.5	24.0	W400
Jharg-ram 1	Tree 16	RRS, Jhargram	8.5	5.0	1.5	NA	W320
Anak-kayam1	BLA 139-1	CRS, Anakkayam	12	6.0	1.7	28	W280
Sulabha	K-10-2	CRS, Anakkayam	23.34	9.8	2.5	25.51	W210
Mridula	PTR1-1	CRS, Anakkayam	3.31	3.6	1.4	38.87	W450
Madak-athara 1	BLA 39-4	CRS, Madakkathara	13.8	6.2	1.6	26.8	W280
Madak-athara 2	NDR 2-1	CRS, Madakkathara	17.0	7.3	2.0	26.2	W280
K22 1	Kottarak-akara 22	CRS, Madakkathara	13.2	6.2	1.6	26.5	W280
Veng-urla 1	Ansur 1	KKV, Vengurla	19.0	6.2	1.6	26.5	W280
Veng-urla 2	WBDC-VI	KKV, Vengurla	24	4.3	1.4	32.0	W320
VRI 1	Vazhisod-hanai palayam	RRS, Vriddhachalam	7.2	5.0	1.4	28.0	W320
VRI 2	Kattupalli	RRS, Vriddhachalam	7.4	5.1	1.4	28.3	W320
VRI 3	Edayan-chavadi	RRS, Vriddhachalam	10.0	7.2	2.1	29.1	W210
Goa 1	Balli-2	ICAR Research complex, Goa	31.5	7.4	2.2	29.8	W210

Source: Peter, 2002

2.4.5 Hybridization

Hybridization techniques for breeding varieties in cashew have been standardized by various workers and 13 hybrids have been developed and released for commercial cultivation in the country. Bhat *et al.* (1998) showed that the paper-roll method of hybridization was better than other methods. In order to combine prolific bearing with other desirable traits like bold nut, cluster bearing habit and compact canopy, hybridization with parents selected for these characters were attempted. Hybrids performed better than selections. Hybrid vigour could easily be commercially utilized in cashew through softwood grafting.

A simple technique of pollination for hybridization involves use of butter paper rolls or pantographic paper rolls which is as follows:

1. Panicle having flower buds that will open the next day are selected both on male and female parental trees. All opened flowers and nuts if any, are removed from selected panicles on female parental tree.
2. Each day between 8 and 9.30 am, all opened flowers from selected panicles on female parental trees are removed. Anthers are removed (emasculated) using ordinary pins before anther dehiscence from freshly opened hermaphrodite flowers of panicle. Stigma along with style is enclosed with a butter paper sheet of 2.5 x 1.5 cm in size by rolling it with help of fingers.
3. Freshly opened male flowers with undehisced anthers are collected in a petri dish from selected male parents and anthers are allowed to dehisce under partial shade.
4. Butter paper roll from emasculated flower is removed and stigma is pollinated with pollen from freshly dehisced anthers of male parent collected in a petri dish.
5. Pollinated stigma along with style is re-enclosed with butter paper roll.
6. Each panicle is labeled indicating parentage as well as panicle number, one panicle being used for only one cross combination.
7. Procedure is continued until 8 to 10 hermaphrodite flower are pollinated in each selected panicle.
8. All opened hermaphrodite flowers that are not used for pollination are removed easily.
9. All remaining flower buds are removed from panicle on last day of pollination.
10. Each panicle with developing hybrid nuts is enclosed in a cloth bag to collect nuts on maturity. Details of crosses should also be written on cloth bags.

Twenty hybrids were released from India of which 15 have kernel grade of W 180 to W 210. These 15 hybrids have at least one of the parents with bold nut character (brazil-18, K-30-1 and Vettore-56) and thus prove usefulness of selecting parents with bold nut character for transmitting this to hybrid (Table.7). Among hybrids released so far, kanaka and priyanka, both of which have BLA 139-1 as a parent, have very short flowering phases. Short duration of flowering, high sex ratio and longer mixed phase, intense branching, high shelling and high nutritive value of kernels are also looked in parents.

VRI (cw) H1 is a high yielding hybrid suitable for cashew growing districts of Tamil Nadu. The hybrid is a cross between M26/2 and M26/1. It has a typical cluster bearing property with 6-10 fruits per panicle. The nuts are bold, kernel 2.2 g, easy peeling testa and high shelling percentage, grade W210. The cashew apple is pink with a yellow tinge. The trees are moderately resistant to tea mosquito bug under field conditions. Aneesarani and Jeeva (2010).

Aneesa rani *et al.* (2011) reported that hybridization and selection experiment was carried out in cashew to evolve a compact or dwarf F1 hybrid suitable for high density planting system, which is the utmost need to improve the productivity of cashew nuts in the country. Eight promising hybrids with tall, semi-tall or dwarf stature were selected from the crosses and evaluated for morphological parameters and stomatal density. The results revealed that the hybrids HC6 was dwarf with 3.0 m height, reduced number of primary and secondary branches, number of leaves per current season's shoot (5.3), reduced internodal length (0.25 cm) and number of internodes (5.3). HC6 also recorded a higher number of stomata (57.2 per mm2). The highest cumulative yield for three years was recorded in HC1 (3.5 kg nuts per tree) followed by HC 6 (2.65 kg nuts per tree). HC 1 was semi-tall and compact. Correlation studies revealed that there is strong positive correlation between plant height and length of current season's shoot, number of internodes per shoot, internodal length and number of branches. This study resulted in a hybrid with reduction in height, number and length of internodes, spur formation on shoots, reduced leaf size, number of shoots with moderate yield. However, the genetic nature of dwarfness is to be confirmed by further evaluation through conventional breeding and molecular approaches.

Performance of cashew varieties under high rainfall zone of Karnataka was studied (Hanamashetty *et al.*, 2002) and encouraging performance of Ullal-1, Ullal-2 and Selection 37 was indicated. Evaluation of cashew varieties under hill zone of Karnataka revealed that Vengurla 6 and Selection 2 performed better. Though Vengurla 6 is less vigorous it excelled others in yield and nut characters (Mahesha *et al.*, 2005).

Table 7. Cashew hybrids released in India and their salient features

Name of Varieties	Percentage	Institutions	Yield Kg/tree	Nut Wt(g)	Kernel Wt(g)	Shelling (%)	Grade
Kanaka (H1598)	BLA 139x H3-13	KAU, Madakkathara	12.80	6.80	2.08	30.58	W 280
Dhana (H1608)	ALGD-1 x K30-1	KAU, Madakkathra	10.66	8.20	2.44	29.80	W 210
Amrutha (H 1597	BLA 139 xH3-13	KAU, Madakkathara	18.35	7.18	2.24	31.58	W 210
Priyanka (H-1591)	BLA 139-1 x K30-1	KAU, Madakkathara	17.03	10.80	2.87	26.57	W 180
VRI (cw) H1	M 26/2 x M 26/1	RRS, TNAUV ridhachalam	14.50	7.20	2.20	30.50	W 210
Vengurla 4	Midnapur Red x Vettur 56	KKV, Vengurla	17.20	7.70	1.91	31.00	W 210
Vengurla 6	Vettur 56 x ansur 1	KKV, Vengurla	13.80	8.00	1.91	28.00	W 210
Vengurla 7	Vengurla 3 x M-10/4	KKV, Vengurla	18.50	10.00	2.90	30.50	W 180
BPP-1	T 1X T 273	CRS, Bapatla	10	5.0	1.3	27.5	W400
BPP-2	T 1X T 273	CRS, Bapatla	11.0	4	1	25.7	W450
BPP-8	T 1X T 273	CRS, Bapatla	14.5	8.2	2.3	29.0	W210

Apple and juice yield potential of 18 varieties were studied (Salam and Joseph, 2003). Variety Kanaka yielded maximum apple (105 kg ha) and juice (70%). Apple size was highest in M 33/2. Varieties kanaka, M 26/2, H 1608, VTH 30, V4, V3, H-1600 and

Ankkayam 1 possess desirable apple and juice characters in addition to nut yield and size and can be recommended for dual purpose.

2.5 Biotechnology

Work on cashew tissue culture has been in progress on developing multiplication protocols from cashew explants. Areas where biotechnology would be of importance in cashew are somatic embryogenesis and plantlet regeneration which could subsequently be useful for genetic transformation to introduce genes for resistance to tea mosquito bug and stem and shoot borers, standardization of micrografting technique, developing haploids and isogenic lines and molecular characterization of existing genetic diversity. Clonal propagation of elite lines, production of clonal rootstocks, *in -vitro* conservation and international germplasm exchange are possible using micropropagation techniques. Micropropagation using seedling explants was standardized. Cotyledon explants produced 10-12 shoots/culture and shoots were rooted by exposing to IBA (20-60 ppm) for 2 hours.

Regeneration from mature tree explants has been difficult due to high rate of contamination, browning, slow growth and poor rooting of micro shoots (Thimmappaiah and Shirly, 2000). Of late, plantlets were regenerated from shoot explants taken from 2-3 year old grafted plants, which were regularly sprayed with Bavistin (0.1%) at weekly intervals. Among different surface sterilization treatments used, a combination of antioxidants, antibiotics and fungicides was better with around 40% contamination free cultures being obtained. MS medium containing full strength macro elements supplemented with Kinetin (5 mg l^{-1}), NAA (1 mg l^{-1}) and Brassinolide (0.1mg l^{-1}) was the best for shoot proliferation and shoot elongation. Micro shoots were rooted *in-vitro* at a frequency of 70-80%, when cultured for 4-8 days in quarter strength MS liquid medium supplemented with IBA (1 mg l^{-1}) for two minutes. Average number of roots ranged from 3-5 with an average length of 1.63 cm. About 50-60% of rooted micro shoots survived and produced healthy and vigorously growing plants, which were transferred successfully to the field (Keshavachandran and Riji, 2005).

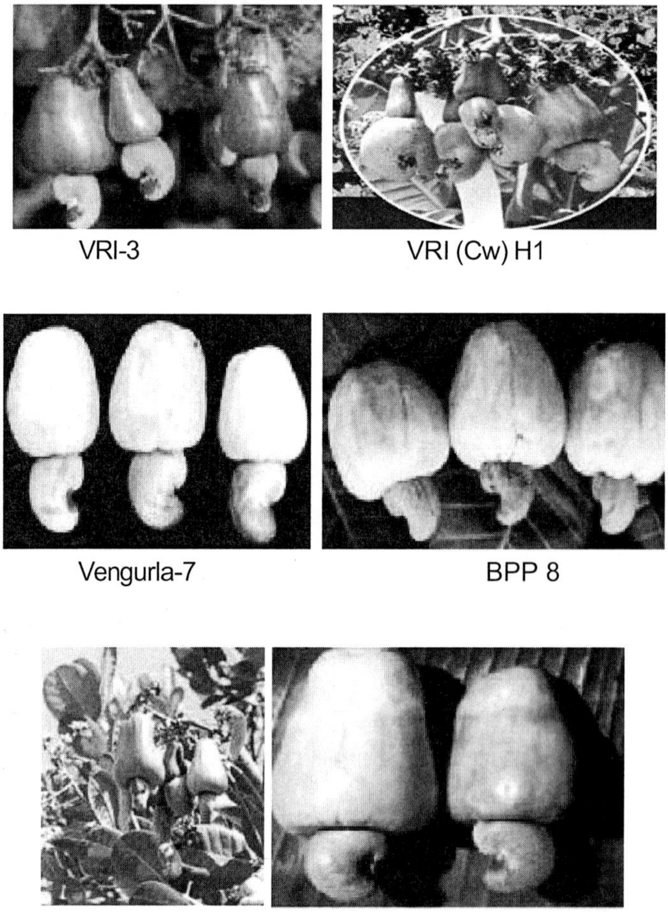

Fig.5. Some important cultivars/hybrids grown in India

2.5.1 Somatic embryogenesis

Somatic embryogenesis in cashew was reported (Anil and Thimmappaiah, 2005). Immature cotyledon segments gave direct somatic embryogenesis. Calluses were initiated from bisected ovules with intact nucelli from 3 weeks old immature cashew nuts cultured on half strength MS medium containing 2,4-D at 1.0 or 4.0 mM alone or a combination of 2,4-D with 12 M Kinetin. Calluses, which became brown after 4-6 weeks, differentiated into somatic embryos on Raj Bhansali medium (RBM) containing 2,4-D (0.01 or 0.5 mM) and 30 mM kinetin. Somatic embryos were initiated in 2.2 - 27.7% of cultures of cultivars Goa 11/6, Kanaka and Ullal-3 only. Number of somatic embryos increased more than 5 folds following suspension culture in RBM basal medium. Somatic embryos matured on half strength MS medium containing 18.9 mM ABA and 60% germinated on liquid medium containing 0.53 m M alpha-naphthalene acetic acid and 4.4 mM 6-benzylamino purine (BAP) giving plantlets with a well developed tap root and a shoot with rudimentary green leaves.

Excised cotyledons from mature embryos of cashew developed adventitious shoots from the proximal end when cultured on MS medium supplemented with 22.2 mM BAP and 3% sucrose. Initiation occurred only at the proximal end of the cotyledon. Differentiation of roots occurred on the same medium after the initiation of leaves. Histology revealed that there was no vascular connection between regenerated shoots and roots. Isolated shoots rooted after treatment on half-strength MS liquid medium containing IBA (122.6 mM) for 72 h. shoots were transferred to half-strength MS solid medium containing IBA (4.9 mM) and IAA (5.7 mM). About 55% of *in-vitro* raised shoots developed roots (Ananthakrishnan *et al.*, 2002).

The embryonic axis and embryonic axis along with cotyledon segments of cashewnut were cultured on woody plant medium supplemented with various concentrations of cytokinins. Maximum number of multiple shoots obtained from embryonic axis with cotyledon segment when cultured on woody plant medium supplemented with 5 mg/ litre (BAP/ AdS), whereas only single shoots were obtained from embryonic axis. Well developed shoots were transferred to rooting medium containing IAA + IBA. The regenerants were successfully transferred to the field (Ananthakrishnan, 1999).

2.5.2 Micrografting

Micrografting technique was standardised in cashew (Mantell *et al*, 1997; Thimmappiah *et al.*, 2002). A successful micro grafting technique to rejuvenate mature cashew tree explants material was developed using in vitro germinated seedling as root stock and shoots from stage II cultures as scions. Grafting success depended on method of grafting, size of scion and age of root stock. Among methods tried, due to ease of operation and faster union, side grafting was preferred. High levels of graft union were obtained when 2-3 cm long shoots from stage II cultures were grafted on to 8-9 days old in vitro raised root stocks in full strength MS medium supplemented with Kinetin (2.5 mg/ l), NAA (1mg/l) and Brassinolide (1mg/l). Very high growth of grafted scion was obtained, when micro grafts were transferred into MS medium containing full strength of macro elements supplemented with Kinetin(5 mg/l),NAA (1mg/l), Brassinolide (1mg/l). Micro grafts after a process of hardening could be successfully transplanted into soil (Keshavachandran and Riji, 2005).

Shoot tips excised from glass house raised seedlings and field grown plants of cashew were micro grafted by a modified side grafting procedure using *in-vitro* raised seedling rootstocks. Factors affecting graft-take were optimized. The micro-scions were obtained with hypocotyls rather than with epicotyls as stock tissue. Rooting of micrografted shoots of mature tree origin was poor (13.3%) because the shoots were only partially rejuvenated (Mneney, 2001).

2.5.3 Somaclonal variation

Somaclonal variation in trees produced *in-vitro* from cotyledonary nodes of cashew was reported (Nivas and D'Souza, 2004). Girth of *in - vitro* produced trees was slightly more and height was lower than that of seed grown trees. Flowers set in earlier in tissue cultured trees. There was a strong tendency towards apical dominance in initial growth of tissue cultured trees. Immune system is also not stable, resulting in a lot of infestation from pests.

2.5.4 Embryo culture

Embryos from immature nuts of cashew were cultured *in- vitro* to regenerate improved hybrid plantlets (Aliyu and Awopetu, 2005). Embryo was excised from developing hybrid immature nuts derived from a diallel cross among 10 accessions harvested at 2,4,6 and 8 weeks after pollination for *in-vitro* culture. Explants were surface sterilized ,aseptically dissected, cultured into pure basal MS medium supplemented with 1 micro M each of NAA, benzyladenine (BA) and GA3 and subsequently observed for germination and survival rates until successful ones were transferred to field. Explants 6 weeks old or older showed better germination rate and highest survival percentage. Germinated explants simultaneously produced shoots and roots and were ready for transfer to field and acclimatization between 90 and 112 days after inoculation. Proplast isolation in cashew has been reported (Rao and Swamy, 2006).

2.5.5 Molecular Markers

In India, application of molecular markers in cashew began only in the year 2000. Karihaloo and Archak (2000) used molecular markers in cashew as an aid to agro-biodiversity management. It was reported that cashew, despite being an introduced plant, exhibits high molecular diversity among its Indian cultivars. This was attributed mainly to the combined effect of its breeding system and continuous seed propagation, that together led to high level of heterozygosity and ensured the availability of multiple alleles. The observed variation in molecular markers, while reflecting the extent of genetic diversity in cashew grown in India, underscored the need to collect and screen the local germplasm for the important economic traits including pest and disease resistance and adaptability to diverse agro-climates.

Characterization of diversity is a prerequisite for exploitation of genetic resources for plant improvement. Presently, characterization of cashew accessions is based largely on morphology (Swamy *et al.* 1998). Although morphological characteristics are still extremely useful, these are often faced with the problems of low penetrance and heritability. Such problems are exaggerated in tree crops because the juvenility period is longer, usually only the maternal parent is known, information on the cultivars other than the released varieties is mainly anecdotal, and local cultivars having different vernacular names may have the same original material. On the other hand, molecular markers are highly heritable, are available in high numbers, and often exhibit enough polymorphism to discriminate closely related genotypes.

Molecular markers such as Random Amplified Polymorphic DNA (RAPD), Random Fragment length polymorphism (RFLP), Amplified Fragment Length Polymorphism (AFLP) and biochemical markers (isozyme, protein) can be used for characterization of germplasm and also for correlating markers with economically important, which will aid in marker-assisted selection. RAPD profiles of 20 Tanzanian accessions of cashew and DNA fingerprints of 34 released and one TMB- resistant accession were reported. Molecular profiles of 24 selections and hybrids were developed using a combination of five RAPD and four ISSR primers pre- selected for maximum discrimination and repeatability. A total of 94 markers were generated which discriminated all varieties with a probability of identical match by chance of 2.8×10^{-11}. There was no correlation between relationships based on molecular data and pedigree of varieties. Narrow range of average similarity values among major cashew breeding centers with only 3.6% of molecular variance partitioned between them was attributed to exchange of genetic material in developing varieties. Difference in average similarly coefficients between selections and hybrids was low indicating need and scope for identification of more parental lines in enhancing effectiveness of hybridization programme (Archak *et al.,* 2003).

About 239 accessions at DCR have been fingerprinted by Thimmappaiah *et al* (2009). Low-level diversity has been observed in 40 elite varieties using RAPD, ISSR and SSR markers at DCR, Puttur. Moderate diversity has been reported in cashew population using protein isoenzyme electrophoretic analysis (Aliyu and Awopetu, 2007; Maranan and Mendiore, 2008). Cashew microsatellites from non-enriched genomic library and sequences of 21 polymorphic SSR markers have been detected and used for multiplex analysis in cashew (Croxford *et al,* 2005). Neto *et al* (1995) showed utility of RAPD markers in distinguishing dwarf seedlings in cashew. Bulk Segregant Analysis (BSA) in germplasm bulks at DCR, Puttur, could identify four RAPD markers linked to economic characters like nut weight and plant stature.

❑❑❑

Chapter 3

Crop Production and Management

3.1 Soil

In Brazil, especially in the north-east, the majority of cashew plantations grow on Quartzarenic Neosols (Quartz Sands), Latosols and Argisols (Podzolics). These are deep soils with good drainage, and with no stones or impervious layers, but with poor chemical fertility (Crisóstomo, 1991). In India, the cashew is cultivated on soil that is infertile, leached, acidic and sometimes containing excess changeable aluminium (Al) (Badrinath *et al.,* 1997). On the other hand, Menon and Sulladmath (1982) reported that satisfactorily prosperous plantations existed on soils that are volcanic, iron-rich, lateritic rusty, alluvial, clayey, and those with a high water table, at times subject to flooding. According to Latis and Chibiliti (1988), the cashew requires fewer nutrients than other fruit trees, and for this reason many plantations are found on soils of marginal fertility. However, research has shown positive responses to mineral fertilizers (Grundon 1999). Falade (1984) concluded that the physical and chemical characteristics of the soil influence the height of the plant as well as the diameter of the crown, and the morphology of the root system. He concluded that light textured soils, free from stones and without an impervious layer or horizon within the top 100 cm are the best for growing cashews. Ideal soil pH is from 5.6-7.0. Soil with a pH more than 8.0 is not suitable for its cultivation, since it is sensitive to excessive salinity. Cashew is sensitive to water-logging and tolerant to moisture stress to a greater extent. Areas prone to water logging may be avoided for its commercial plantations. Before establishing a plantation, the soil should be sampled and analysed to determine the need for soil amendments and fertilizers. In already established orchards, soil and leaf analysis provide supplementary information as to the recommendations for fertilizers and soil amendments.

3.2 Climate

The cashew tree is an evergreen plant, although a partial replacement of the leaves can occur. Owing to its sensitivity to low temperatures, it geographic distribution is confined to regions between 27°N (South Florida) and 28°S (South Africa) (Frota and Parente, 1995). Inspite of being a tropical fruit in origin, the cashew develops well in temperatures varying from 22 to 40°C, although Parente *et al.* (1972) cite 27°C as the ideal average temperature for normal development and fruit bearing. Lower temperature at high altitudes affects development of tree and flowering. Maturity is delayed and nuts are spoiled by early rains. A distinct dry period of 4-6 months with annual rainfall of 1000-2000 mm is ideal for commercial cultivation. Rain during flowering and harvesting is not desirable and causes tea mosquito and *Colletotrichum* infection. In areas with low rainfall, cashew requires irrigation, provided there is sufficient water supply and humidity is not critical.

Owing to the influence of altitude on temperature, cashew plantations may be found at altitudes up to 1,000 m close to the equator. In higher latitudes and altitudes above 170 m, the yield is negatively affected (Aguiar and Costa, 2002). The cashew develops well between 70% and 85% relative humidity. Trees will grow in regions where the relative humidity is 50% for a long period of time, if the soil contains a good reserve of moisture or irrigation is used. In regions where the air relative humidity is above 85%, fungal diseases of the leaves, flowers and fruits increase.

Cashew is a sun loving tree and does not tolerate excessive shade. Therefore, it is necessary to clear the site selected of all forest growth to avoid excessive shade. Average sunshine of nine hours per day was optimum for best production. Wind has little influence on a cashew plantation. However, at velocities of 7 m/s or higher, Aguiar and Costa (2002) reported an increase in loss of flowers and fruits and trees being blown over. In plantations that are extremely close to sea, high wind speed may lead to scorching of buds in new flushes. In such areas, it will be necessary to raise fast growing perennials as windbreaks around plantations. This will not only protect cashew plants from unfavourable hot wind, but also will provide sufficient relative humidity through transpiration.

3.3 Site selection

Selection of a suitable site is very important for its cultivation. Mahapatra and Bhujan (1974) suggested a rating chart for land selection for cashew. Instead of considering type of soil alone, class of a soil with a grading from Class I to Class V should be adopted while selecting site for raising cashew orchards. Class I to Class III types of soil with medium acidic range to neutral (6.3 to 7.3 pH), with a slope of 0 to 15^0, and with water tables up to 10 m were recommended as the best soils suited for higher production of cashew. However, in many countries, other plantation crops such as coconut and rubber compete for similar types of lands, and hence only Class IV and Class V soils are at present committed to cashew. Class V soils that are unsuitable for good production of cashew should be avoided, whereas Class IV soils require strong soil conservation measures as well as other soil amelioration technologies (Table 8).

3.4 Preparation of Land

It is better to select the land having good drainage and devoid of sub-surface hard rock or hard pan for successful cultivation of cashew. When selecting land for cashew, soils with salinity/alkalinity or water-logging should be avoided. Soil depth, slope, course texture, soil fertility and water availability seem to impose very little limitations as cashew is a hardy crop. The land should be ploughed thoroughly and levelled in case of agricultural lands. In case of forest lands, the jungle should be cleared well in advance and the debris burnt. After clearing the jungles, land is to be terraced or bunds constructed in slopy land. Pits of 1 m^3 are to be dug and allowed to wither.

3.5 Propagation

Cashew can be propagated by seeds as well as by vegetative means.

3.5.1 Seed propagation

Since cashew is a cross pollinated crop, its seedling progeny is highly variable. Seed propagation is seldom practiced now except to raise the rootstock materials. Seed nuts collected from selected mother plants of 15-25 years old with intensive branching habit and high yield potential. Seeds should be collected during the month of March to May. Mature, medium-sized nuts which sink in water are collected and dried in sun for 2-3 days. Seed nuts are soaked in water for 18 -24 hours and then sown at 4 cm depth in polythene bags of 20 cm x 15 cm with stalk end up during May gave better germination and growth than cracked seeds or those sown at 8 or 12 cm (Amoah, 2005). Seeds germinate in 7-10 days and seedlings are ready for transplanting in

2-3 months. Cashew seeds when kept in moist gunnies for 7 days or treated with sodium hydroxide 10 or Gibberelic acid 100 ppm recorded significantly high rate of germination and seedling growth. It is likely that these treatments removed the waxy layer of the pericarp thereby facilitating germination and seedling growth (Harris *et al.*, 1994).

The effect of nut weight on germination rate and subsequent seedling vigour of cashew was studied using current season nuts that were stored in to 3 different weight glasses of small (3-<6 g); medium (6-<8 g); and large (8-10 g). Large and medium size nuts performed better than the small nuts in terms of germination percentage, overall seedling vigour and other growth parameters evaluated. Current season nuts weighing between 6.0 and 10.0g are recommended as good quality planting materials for cashew farmers. Such nuts germinate uniformly and rapidly, producing vigorous seedlings that ensure the successful establishment of the young tree (Adebola, 1999). Aneesarani et al. (2004) reported that application of 2g Azatobacter, P$B and 10g Arbuscular Mycorrhiza gave maximum height, girth, root length, number of roots/root fibers in cashew seedings. The study resulted in early maturity and advancement of seedling growth for grafting.

Table 8. Guidelines for selection of land for cashew

Soil characteristics	Very good Class I	Good Class II	Fair Class III	Poor Class IV	Unsuitable Class V
Depth of soil	>1.5 m	90cm -1 m	45cm - 90cm	23cm - 45 cm	<23 cm
Texture	Loam Sandy loam	Loamy sand Silty loam Coastal sand	Clay loam Silty loam Sandy clay loam	Gravelly clay loam Gravelly silty loam Gravelly sandy loam	Gravelly clay Sandy clay Silty clay Clay
Reaction	Very slightly acidic to neutral (pH 6.3 to 7.3)	Slightly acidic (pH 6 to 6.3)	Medium acidic (pH 5.6 to 5.9)	Strongly acidic (pH 5.1 to 5.5) or mildly alkaline (pH 7.4 to 7.8)	Very strongly acidic (pH less than 5) or alkaline (pH more than 7.8)
Land features	Very good Class I	Good Class II	Fair Class III	Poor Class IV	Unsuitable Class V

Slope (%)	<3	3 - 5	5 -15	15 -25	>25
Water table (m)	2-5	1.5 -2 (coastal belt)	8 -10	10 -13	>13
Erosion condition	None to slight (e_0)	Slight (e_1) (sheet erosion)	Moderate(e_2) (rill and sheet erosion)	Severe (e_3) (gully erosion)	Very severe (e_4) (gully and ravine erosion)
Drainage	Well draine d	Well drained to somewhat excessively drained	Moderately well drained	Excessively and imperfectly drained	Poorly drained
Physiography	Coastal plains	Alluvial plain	Plateaus	Denuded hill slopes with shallow soils	Swamps
	Delta reache s	Natural levees	Hills	Ridges	Valley bottoms
Climate and Environmental factors	Very good Class I	Good Class II	Fair Class III	Poor Class IV	Unsuitable Class V
Altitude(m)	<20	20 -120	120 - 450	450 -750	750
Rainfall(cm/yr)	150-250	130 -150	110 -130	90 -110	<250
Proximity to sea (miles)	<50	50 -100	100 -150	150 -200	<200
Temperature (F^0)					
Max. in summer	90-100	100 - 103	103 -106	106 -110	<110
Min. in winter	60	57- 60	53 - 56	48 -52	<48
Humidity (%)	70 -80	65 -70	60-65	50 - 60	<50 or >80
Occurrence of frost	None (once in 20 years)	None (once in 15 years)	Very rare (once in 10 years)	Occasion al (once in 5 Years)	Very often to frequent (every year)

Source: Mahapatra and Bhujan,1974.

3.5.2 Vegetative propagation

As cashew is a cross-pollinated crop, propagation by vegetative means was attempted. Different methods of vegetative propagation - air-layering, veneer grafting, epicotyl grafting, softwood grafting and *in situ* grafting were tried in cashew with varying success. Among the various methods tested, softwood grafting (Fig.6) was found to be the best for vegetative propagation (Swamy *et al*, 1993). Of late, softwood grafting has become the most successful and widely adopted method. It has also been shown that softwood grafting is feasible for commercial multiplication. Based on these results, India has been producing over 15 million grafts annually under both government and private sectors.

Air-layering was found to be one of the popular methods among growers. This technique however, produced trees with poor anchorage as the root density was found to be low. It also resulted in poor field establishment and high susceptibility to cyclones and drought conditions. Air layering was therefore found to be unsuitable for commercial exploitation.

In the case of mound layering too, the absence of a tap root was found to be a disadvantage. Epicotyl grafting was another method that had limitations due to high mortality at transplanting and incidence of collar-rot at the nursery stage. Soft-wood grafting developed at the research centers was found to be the most viable method of propagation that was commercially acceptable; it gave a success rate of about 70 per cent. The technique of soft-wood grafting described below is similar to epicotyl grafting except the difference in the age of the rootstock.

Forty to sixty day old seedlings are used as rootstocks. Two pairs of leaves are retained and the seedlings are decapitated at the soft-wood apical region. Wedge-grafting is then carried out with a 4-5 cm cleft on the rootstock and with a small portion of the inner surface removed to facilitate a perfect union of the wedge -shaped scion, which has been prepared by shaving a portion of the bark and tissue on either side. The union is then secured by tying with a 15-30 cm polythene strip. The top of the scion is covered with a polythene cap to protect the apical portion of the scion from desiccation. The grafted plants are maintained in a low cost mist chamber or screen house for 8-10 days until sprouts emerge, and then the grafted plants are provided more sunlight and the caps removed. This wedge grafting technique is carried out by using only the soft-wood tissues of the stock and scion. Successful grafts show signs of growth 3-4 weeks after grafting. They become ready for planting 5-6 months after grafting. Ideal season is from March to September. The following management techniques are important in nursing young grafted plants.

Kadam *et al.* (1996) revealed that *in-situ* softwood grafting is a successful method of clonal propagation in cashew and could be used for achieving large scale multiplication of high yielding plantations of improved cashew varieties. The percentage of graft success was maximum with the budsticks of 120 days age which was at par with the budsticks of 150 days age. The age of budsticks as well as size influenced the graft success significantly. Highest percentage of grafting success was obtained with scion sticks of 12.5 cm length followed by scion sticks of 10.0 cm length. The growth of the successful *in-situ* grafts was also vigorous and encouraging. The higher grafting success of 73.33 to 83.33 per cent and also survival of the grafts were obtained between the period from 15th July to 15th September; the month of rainy season.

Grafts need to be watered frequently depending on the season. Excess water needs to be drained by providing drainage holes in polybags. Shoots on the rootstocks have to be nipped off frequently. Polythene wrapping at the union has to be removed about three months after grafting to prevent girdling. When the scion leaves turn from brown to green, rootstock leaves have to be removed (approximately 60 days after grafting). Flower shoots that sprout during the normal flowering season should be removed at the nursery stage. To prevent roots penetrating into the ground, grafted plants should be shifted frequently or placed on thick gauge black polythene sheets. Partial shade has to be provided to avoid sun-scorch by placing the grafted plants in a low cost mist chamber/screen house. Direct sunlight should be avoided as polythene bags tend to perish. Watering on daily basis should be done in summer. Regular insecticide sprays need to be given to control leaf sucking insects. When transporting grafted plants, terminal shoots and taproots should be protected.

In cashew, the size of nuts sown had been reported to have significant influence on the morphological growth performance of the seedlings in the nursery (Adebola, *et al.*, 1999) and they therefore recommended the use of bigger nut sizes rather than smaller nut sizes for raising cashew seedlings for plantation establishment. Their findings were based on morphological growth measurements of the cashew seedlings rather than the physiological growth rates. These measurements are not as efficient as growth analysis in estimating the crop growth since they do not measure the dry matter changes over time and neither do they measure net growth rate of crops. Besides, Hammed (2008), found that the size of cashew nuts sown, do not only influence the morphological growth performance of seedlings, it also affects dry matter yield in favour of bigger nut size. Thus, understanding of the relationship between dry matter yield and leaf area of cashew seedlings with respect to nut size, will aid understanding of processes and forms within the seedling for the overall development of the crop.

Influence of anthracnose disease of cashew seedlings on graft success in cashew nurseries was recorded (Venkateswarlu *etal.,* 2003). A progressive decline in number of successful grafts was observed with increase in disease severity of rootstocks. Grafts registered significant decreases in elongation of graft and mean leaf area over control irrespective of the extent of disease severity. Drenching of Bavistin 50 WP once in a week to prevent root rot is essential streptocycline drenching would help mitigate bacterial rot in seedlings.

Flushing grafting wherein grafting is done almost throughout year using post-harvest, pre-harvest and induced flushes of 21 days old on rootstocks of 21 days are also becoming popular. Success percentage is high when kept in mist chamber.

3.6 Soil preparation

For new plantations, existing surface vegetation and roots, especially around the area where the pit will be prepared, should be removed to minimise competition from other plants. For soil preparation, the use of heavy machinery should be avoided, to diminish the risk of soil compaction.

Liming, when necessary, should be performed in two steps, the first before aeration and the second at ploughing. The quantities should be sufficient to increase base saturation to 60% and the levels of exchangeable calcium (Ca) and magnesium (Mg) to a minimum of 3 and 4 mmolc/dm3, respectively (Crisóstomo *et al.,* 2003). Chisel plough would loosen the soil for better aeration and root growth.

Rootstocks

Scions

Grafting Grafted plant

Mist chamber Grafts ready for planting

Fig. 6. Softwood grafting of cashew

3.7 Field planting

Softwood grafts of high-yielding varieties are the best planting materials, although seedlings are used for planting. Pit size influences the growth of plants. Different size of pits are followed in cashew growing states 45x45x45 cm (Tamil Nadu), 90x90x90 cm (Kerala) and 60x60x60 cm(Karnataka) in light to medium soils. If a hard substrate like laterite is present, pits may be 1m x 1m to compensate for the lesser depth of soil. It is preferable to dig pits 15-20 days before planting to expose planting holes to direct sunlight which can help remove termites and other harmful insects that can damage young plants, if present. When filling, top soil mixed with compost (5 kg) or poultry manure (2 kg) and 200g of rock phosphate are placed in the pits. Contour planting is usually followed in sloping areas. Standard conservation measures need to be followed on steep lands when establishing cashew plantations.

Young plants are planted in the months of July-August. Three month old grafts are suitable for planting in main field. Most nurseries supply 5-12 month old grafted plants in polybags. At planting, the polythene bag is removed without disturbing the ball of earth and the roots. Care is taken to place the grafted plant in the pits leaving the graft joint at least 5 cm above ground level. Normally the scion is staked to avoid damage from wind and the support should remain up to third year from planting. Most orchard growers use mulch around the planting hole to suppress weeds and conserve moisture.

3.8 Spacing

Spacing varies depending on soil type, agro-climatic conditions and management methods. The spacing recommended is 7.5 m x 7.5 m for poor soils and 10 m x 10 m for deep rich soils and sandy coastal area. A spacing of 7.5m x 7.5m or 8m x 8m is recommended for cashew which gives a tree density of 175 and 156 trees per ha, respectively. In level sites, however, it would be advantageous to plant cashew at a spacing of 10m x 5m which will give a tree density of 200 trees per ha and at the same time providing sufficient space for growers to plant inter-crops during the initial years of establishment. On very sloppy areas, rows may be spaced 10-15 m apart with spacing of 6-8 m between trees in a row. Square or triangular system of planting can be adopted. Triangular system accommodates more number of plants/ha. June-July or September or October is ideal for planting.

An experiment was carried out in the red laterite tract of West Bengal to study the effect of different plant spacings viz; 5 x 5m, 6 x 6 m, 8 x 8 m, 10 x 10 m, 6 x 6 x 6 m, 8 x 8 x 8 m, 10 x 10 x 10 m and 10 x 5 m on yield of cashew cv. Red Hazari. The highest yield was recorded from the plants spaced at 8 x 8 m throughout the period of investigation followed by 10 x 5 m rectangular planting. The unit area yield had markedly increased with the increase in plant density and it was highest at a plant density of 400 plants/ha (5 x 5m) followed by 321 plants/ha (6 x 6 x 6m) and minimum at 100 plants/ha (10 x 10m) (Ghosh *et al.*, 2000).

3.9 High density planting system

High density planting enhances productivity of cashew considerably through better utilization of environmental resources. High-density planting system in cashew doubled nut-yield during the first 10 years of planting (Yadukumar *et al*, 2002). Selection of right variety and its planting density are prerequisites. This involves planting of more number of grafts per unit area initially and thinning out at later stages (after 8-10 years) depending upon the canopy expansion rate. Instead of 177 plants/ha (7.5 m x 7.5 m), 312-625 plants/ha are planted (8 m x 8 m, 4 m x 4 m). In Tamil Nadu, this closer planting technology at 5 x 4 m to accommodate 500 plants ha^{-1} was standardized and released for adoption by TNAU during 2011 (Aneesarani *et al.*, 2011) (Anon, 2011). A yield of 3000 kg nuts/ha could be harvested at the age of six years. High Density Planting recommended to farmers of Tamil Nadu by organizing seminars, workshops, field days, campaigns training programme, exposure visits every year. This technique has been promoted by conducting frontline technology demonstration programs through TANHODA of Govt. of Tamil Nadu under this Hi-tech HDP programme, 1000 ha was brought under HDP during 2010-14. All the new area expansion programmes were recommended under HDP system.

Considering soil fertility status, climate, management practices and initial plant population is to be decided carefully for every agro-climatic condition. If soil is fertile 5 x 5 m spacing (400 plants/ha) can be maintained to get the highest yield and income. In poor fertility areas, 4 x 4 m (625 plants/ha) is recommended (Yadukumar *et al.*, 2003).

Further, based on the physiological response, optimum tree population (625 trees /ha. at 4m x 4m spacing) for high density plantations for cashew was suggested (Balasimha and Yadukumar, 1993). This was considered as threshold level for maximizing canopy surface area, photosynthetic capacity and yield in cashew on an unit basis, although, net photosynthesis and transpiration values were significantly higher in wider spaced trees that received higher irradiance on an individual basis. Tree competition for irradiance was observed to be an important factor and hence, high density cropping system should aim at optimizing this limiting factor as reported earlier.

High density planting is more suitable for soils with low fertility. Further, high density planting reduces weed growth due to early ground coverage by the crop canopy and also reduction in soil temperature thereby increase in soil moisture content especially during peak summer season and also through mulching effect. Regular pruning is needed to contain the canopy. In later years for instance, eleventh year after planting, it is necessary to thin out the tree population to 50 per cent by removing every alternate tree in each row. Cashew yield can be increased by 2 to 3 folds by high density planting as compared to the normal density planting. During later years, plant population is to be regulated by selective felling to minimize competition. This technique is more useful in poor soils where rate of expansion is slow.

When plants attain full growth, spacing between plants can be regulated to 8 m x 8 m. During early years, though per tree nut yield is more or less same with all trees, irrespective of density of planting, per hectare yield is more from high density planting. Large quantities of firewood can be obtained while cutting excess trees, which fetch additional revenue. Weed growth in such plantation can also be checked effectively. Relatively short-statured and compact varieties such as Anakkayam-1 and Dhana are more suitable for high density planting, since thinning operation can be delayed to more than 10 years. Salam (1999) reported that the varieties 'M 44/3', 'Anakkayam-1' and 'H-1608' were suitable for high-density planting. A dwarf hybrid HC 6 was evolved at Regional Research Station, Vriddhachalam, which is highly suitable for high density planting system. It's due to dwarf hybrid with a smaller round compact canopy, 3.0 m tall, less spreading, intensive branching, smaller trunk girth, smaller leaves, lower number of internodes, reduced internodal length and higher stomatal density (Aneesarani *et al.,* 2011).

VRI-3, which is very popular among Tamil Nadu farmers due to its export grade kernels is highly amenable for pruning and hence suitable for HDP system. The hybrid VRI(cw)H1 produces enormous number of current season shoot, flowering shoots after pruning and as a result high nut yield. The high density planting spacing 5mx4m (Fig. 7a and 7b) was released as technology and followed in new planting programmes in Tamil Nadu.

Fig. 7a. High density planting (5m x 4m) field at Regional Research Station, Vridhachalam

Fig. 7b. High density planting demonstration plot at farmers field in Cuddalore district

3.10 Ultra-high density planting system in cashew

3.10.1 Introduction

With the shrinking cultivable land area for cultivation of fruits and vegetables, the availability of land for a drought tolerant and hardy crop is reducing greatly at a faster rate. Planting the cashew trees at a closer spacing would accommodate more number of plants per unit area, more photosynthetic area for bearing and thus, higher yield per unit area/available land. With the current practice of growing cashew seedlings in a haphazard manner to accommodate hardly 60 plants per hectare, farmers are unable to achieve the national highest average yield of 1500 kg /ha. The present recommendation of 7x7 m would accommodate 200 plants per hectare yielding higher to the tune of 1000 kg/ha wherever farmers adopted apt management technologies in right time. Ultra-high Density planting system in cashew with 3x2m, 4x2m and 5x4m would be a new venture and if managed along with proper drip and fertigation system high soil fertility, it would be the most precise and state - of the - art technology to increase the productivity of cashew in Tamil Nadu and on the whole in India.

3.10.2 Scope

High density planting is a modern method of fruit cultivation involving planting of fruit trees densely, using trees/varieties which are modified for canopy for better light interception, precocious cropping and regular high yields of good quality fruits. As the value of land is increasing, the available land for cultivation is decreasing; high density plantation is a very important and effective venture as great amount of fruit bearing volume per hectare would be attained at early years. Higher planting density produces higher level of photosynthetic active radiation with high leaf area index and better utilization of the available solar radiation.

Based on the recommendations of DDG, ICAR and the necessity of closer planting technologies with canopy management strategies, a new special grant project has been acquired by the scientists of TNAU and implemented during 2010-11 at Regional Research Station, Vriddhachalam.

3.10.3 Ultra High Density Planting System

The technology "Ultra-high density planting system in cashew" consisted of planting 1200 to 1666 plants per hectare combined with canopy management, foliar spray and drip fertigation to double/triple the nut yield. A novel technology has been standardized in cashew by planting grafts at 4 x 2 m (Fig. 8a and 8b) or 3 x 2 m to accommodate 1200 plants and 1666 plants per hectare respectively. Training of plants upto six months after planting to contain in the provided space was essential. Training consisted of removal of sprouts in the grafts upto 75 cm from the ground level, followed by formative pruning to develop 2-3 primary branches, 4-5 secondary branches and 5-6 tertiary branches/secondary branch. The plants put forth flowering at 12 months of planting. (Aneesarani and Keisarlourdusamy, 2016).

A nut yield of 250 kg/ha recorded from one year old plants. The plants were pruned during second week of July (tertiary branch pruning) to induce current season's shoot production. Drip irrigation was given daily @ 8 l/day for first one month, 16 l/day at alternate days from 6-8 months, 32 l/day once in 4 days from 8-12 months (except rainy days). Liquid fertilizers given in two splits (July and November) to supplement the recommended dose of 225:75:75 kg/ha. Foliar spray schedule was given as per recommendation and treatments. Increase in number of current season's shoots and number of flower panicles has doubled the yield. Actual yield per unit area recorded from two year old plants was 170 kg/0.35 acre in 3 x 2 m and 200 kg /0.35 acre in 4 x 2 m (compared to 81.2 kg/0.35 acre in 7 x 7 m and 112.5 kg/0.35 acre in 5 x 4 m). Estimated yield was 1275 kg/ha in 3 x 2 m, 1500 kg/ha in 4 x 2 m (compared to 609 kg/ha in 7 x 7 m).

3.10.4 New findings from the experiment

Canopy management started from one month after planting. Basic framework of the plants was done within six months after planting with defined number of primary, secondary and tertiary branches. The spacing 4 x 2 m was the best for canopy management, light interception, more number of current season's shoots, flowering panicles, fruit set and yield. Due to proper canopy management, the trees were brought to flowering and yield from first year after planting itself. A yield of 1520 kg nuts per hectare recorded in 4 x 2 m spacing at 2nd year after planting itself. The technology would be a great boon to the cashew farmers in the country for high remuneration from first year itself. (Aneesarani and Keisarlourdusamy, 2016).

UHDP planting (4m x 2m spacing)

Intercropping with black gram

Six months old cashew in UHDP

Tipping of apical shoot to induce axillary shoots

Training of the plant for canopy architecture

UHDP System in Cashew at 4 x 2 m

Fig. 8a. Ultra High Density Planting at Regional Research Station, Vriddhachalam

Two years old plants

Plants ready for pruning during July

Pruned field

Flowering and nut set after pruning

Flowering in 5 years old VRI-3 trees

Bearing in Ultra High Density Planting

Fig. 8b. UHDP – Cashew (contd.)

3.11 Pruning and Foliar Spray for increasing yield in Cashew

Cashew is a perennial tree crop bearing on its current season's shoot. At Regional Research Station, Vriddhachalam, pruning experiments were conducted for six years along with foliar nutrient spray. A technology has been standardized on pruning and foliar spray schedule to increase the current season shoot, flowering and yield in cashew. The technology consisted of pruning tertiary branches every year during June-July to induce more number of current season shoots followed by a foliar spray schedule. The foliar spray schedule consist of NPK 19:19:19 @1% at new flush stage (August), mono-ammonium phosphate @ 1%+boron 0.1% at flowering stage (December) followed by third spray of TNAU Panchagavya (An organic formulation of five ingredients from domestic cow) 3% at fruit set stage (January-February). Recommended fertilizer dose and plant protection measures were given.

It has resulted in increased number of current season's shoot (Fig. 9), shoot length, number of leaves per shoot, leaf wet weight, leaf dry weight, leaf area, number of flower clusters per sq.m, nut weight and yield. The leaf nutrient content and quality of kernels were also improved by the pruning and foliar spray treatment. This technology has resulted in a yield increase by 42 per cent. BCR of 5.3 was obtained from this technology. Partial budgeting analysis has revealed an additional return of Rs. 11,640 per acre over and above the control due to the adoption of this technology. On-farm trials conducted in farmer's field resulted in 36.5 per cent yield increase over the previous years.

Fig. 9. Vegetative growth and fruit set in pruning and foliar sprayed field

3.12 Manures and Fertilizers

Cashew is often grown on marginal soils and on wastelands mostly unsuitable for other economic crops. Nitrogen and P were found to be the most important nutrients during pre-bearing stage. At the bearing stage, K, together with N, is important. Application of fertilizers, their dosage, time and schedule under different agroclimatic zones has been standardized (Yadukumar *et al*, 2008).

Due to extensive root system, cashew draws nutrients from large volumes of soil and performs well on poor soils where other crops fail. However, it shows good response to fertilizer application. Application of 10-15 kg of farmyard manure or compost annually is generally recommended for cashew. In addition, the current fertilizer recommendation is 500g N (1.1 kg urea), 125g P_2O_5(625g rock phosphate) and 125g K_2O (208g muriate of potash) per tree per year. This has increased production in the All India cashew trials carried out at the research centers. These trials also showed that the cashew responds well to increased N applications upto 750g. This dose may be increased by 33% in areas where soil fertility is low. Since local NPK fertilizer mixtures do not deliver the required nutrients, application of straight fertilizer is recommended.

Fertilizer is applied annually at the end of the rainy season into a shallow trench at the drip line of trees. It is also recommended that fertilizer be applied in split doses during pre-monsoon (May-June) and post-monsoon (September-October) periods to assure better uptake of nutrients. If a single application is done, the post-monsoon period is more suitable when ground moisture is adequate. One third the recommended dose is applied in the first year, two third the dose in the second year and the full dose thereafter (Table.8). For young trees, fertilizers over entire tree basin 10 cm deep, within canopy. For adult trees, fertilizers should be applied in 15 cm basins 50 cm away from tree trunk. Since, cashew is a surface feeder, 75% feeder roots are seen within 30 cm depth and active root zone of cashew resides within 2 m radius around tree applying fertilizer within this area is the most effective for utilization of applied nutrients. The highest root activity and peak absorption of nutrients take place during flushing and early flowering phase, which extend from September to December. The fertilizer application is restricted to once a year coinciding with flushing and flowering.

Rootstocks and grafts of cashew supplied with 150:20:100 ppm of NPK at the rate of 100 ml/plant/week had higher plant-height, stem-girth and number of leaves (Manjunatha, 2001). The ratio 2:1 of N:P is ideal for young cashew trees (Shi-Wenge *et al.*, 2005). Application of mineral nutrients in combination with organic fertilizers significantly increased plant height, dry weight of aerial parts and number of leaves in cashew seedlings (Lima *et al.*, 2001).

A 30-years old cashew tree removes 2.85 kg of N, 0.75 kg of P2O5 and 1.27 kg of K2O (Mohapatra *et al.*,1973) from the soil. Leaf N content of 1.51% in the month of April is considered optimum for high nut yield (Ghosh and Bose, 1986). Macronutrient removal by cashew nuts and the cashew apple was N > K > Mg > P > S > Ca and K > N > Mg > P > S > Ca, respectively (Fragoso, 1999). Use-efficiency of N and K in cashew was 24.7% and 12.37%, respectively (Latha and Salam, 2001).

Table 8. Recommended manurial schedule is as follows:

Age of plantation	Manures per tree			
	FYM or Compost (Kg)	N (g)	P (g)	K (g)
One year old	10	50	25	25
Two years old	20	100	50	50
Three years old	20	150	75	75
Four years old	30	150	75	75
Five years and above	50	500	125	125

Table 9. Recommended Doses of NPK Fertilizer for Cashew (g/plant)

Year	Urea (gm)	Rock Phosphate (gm)	Muriate of Potash
1	330	200	70
2	660	400	140
3 onwards	1,100	625	208

Based on the results of research conducted by the National Coordinated trials, the following methods of fertilizer application are recommended to cashew growers. In the red loamy soils in low rainfall areas such as the east coast, fertilizers have to be applied and raked into the soil along the drip line of tree canopies. In laterite soils and steep lands of the west coast, fertilizer is applied in circular trenches of about 25 cm width along the drip line of trees. Trenches are filled and mulching around the tree is done to ensure soil moisture retention.

One of the major reasons for low productivity of cashew in Tamil Nadu as compared with other states like Andhra Pradesh and Maharashtra is poor nutrient management practices in orchards. However, studies proved that the cashew responds very well to fertilizer and irrigation, wherein, the individual trees are capable of yielding 23 -36 kg per tree (Chacko, 1994).

There is a great scope for improving the fruit set, number of nuts and nut yield per tree through fertilization, foliar spray with nutrients and plant growth regulators as evidenced in many other horticultural crops (Rajangam *et al.*, 2001; Albert and Krishnasamy, 2004). The problem of declining productivity has led to an investigation into the effectiveness of nutrients as foliar application to improve the fruit set and fruit retention in cashew.

Foliar sprays of nutrients (urea 2 to 4%; DAP 1%; orthophosphoric acid; ZnSO4 4%; Cu 0.3 to 0.6%), at emergence of the flush, panicle initiation and fruit-set stages, ensure better fruit-set and enhanced nut yield (Ghosh, 1988; Sapkal, 2000). Yellow leaf spot occurrence in low soil pH (4.5–5.0) could be corrected by foliar sprays of Mo salts (Subbaiah *et al.*, 1986).

The technology consisted of pruning tertiary branches every year during June-July to induce more number of current season's shoots followed by a foliar spray schedule. The foliar spray schedule consisted of NPK 19:19:19 @ 1% at new flush stage(August), mono-ammonium phosphate @ 1% + boron 0.1% during flowering (December) followed by third spray of TNAU panchagavya 3% during fruit set stage (January-February). This technology resulted in higher nut 30 % increase in yield compared to control (4.0 kg/tree) (Aneesarani *et al.*, 2011).

Aneesarani *et al.* (2011) reported that foliar spray of NPK 19:19:19 @ 1 per cent and MAP 1 per cent at flushing and flowering stage respectively enhanced the bisexual flowers in VRI 3 cashew. Murali (2012) revealed that the treatment (M_1S_8) consisting of tertiary pruning along with two foliar sprays *viz.*, NPK 19:19:19 @ one per cent at new flush stage (August) followed by a second spray of boron 0.1% + SOP 2% + MAP 1% at flowering stage (December) was found to be the best in increasing nut yield, yield attributing characteristics and quality namely number of panicles, percentage of bisexual flowers, percentage of fruit set, nut weight, kernel weight, kernel protein, kernel fat and kernel calcium in cashew grown under high density planting system (Murali *et al.,* 2015).

3.13 Functions and importance of nutrients

3.13.1 Nitrogen (N)

Reddy *et al.* (1981) and Ghosh and Bose (1986) report impressive increases in production by increasing the amount of nitrogen applied. Ghosh (1989) showed that increasing the amount of N applied significantly increased the duration of flowering and the number and weight of nuts. The symptoms of N deficiency are seen initially in the older leaves, characterized by chlorosis in the apex region of the lamina, but because N can be mobilized and redistributed within the plant, the young leaves remain green. Generally, plants deficient in N have: (a) low stature, fewer branches and fewer leaves; (b) pale leaves due to less chlorophyll; (c) when deficiency is severe, leaves fall and branches die. Chemical analysis of the leaves together with soil analysis is used to evaluate the nutritional state of the orchard and formulate fertilizer recommendations.

3.13.2 Phosphorus (P)

Phosphorus is required in a smaller amount than either N or K. The quantity of P exported by the fruit ranging from 0.7 to 4.15 g/kg and the pseudo fruit from 0.11 to 1.30 g/kg. Such differences are due to the form of expression of the results, fresh weight and dry weight. The visual symptoms of P deficiency are characterized, initially, by dark green coloration of the leaf, which in the more advanced stages, turns an opaque green before the leaves fall prematurely. In general, plants deficient in P have smaller leaves than well-nourished plants. Due to nutrient redistribution, the visual symptoms are observed in the leaves on the lower third of the crown.

3.13.3 Potassium (K)

At harvest, 1000 kg cashew nut and 10,000 kg of fresh peduncle (cashew apple) contains about 115 kg K (Fragoso, 1999). The symptom of K deficiency is similar to that of N and P, starting in the oldest leaves, which show light chlorosis on the edges. In the advanced stages, the chlorosis reaches the centre of the leaf lamina, remaining green only at the base, in appearance like an inverted "V". Remembering that visual symptoms only become evident when the deficiency is in an advanced stage, chemical analysis permits a more accurate diagnosis. Kernot (1998) considers an adequate leaf K is between 7.2 and 11.0 g/kg. On the other hand, the values found by Haag *et al.* (1975) are much larger and vary between 11 and 20 g/kg. This difference, possibly, may be attributed to the genetic material used because the latter author worked with the Giant Cashew.

3.13.4 Calcium (Ca)

The initial symptoms of Ca deficiency, according to Avilán (1971), are seen as ripples in new leaves. Because Ca has little mobility in the plant, it has to be applied frequently.

3.13.5 Magnesium (Mg)

The quantity of Mg absorbed by the cashew plant is generally less than that of Ca and K. In general, Mg deficiency is due to competition with other ions like Ca2+, K+ and NH4 + for uptake by roots (Mengel and Kirkby, 1978). Most of the Mg in the plant is in chlorophyll but it is also an enzyme activator, especially those involved in the transfer of phosphate radicals rich in energy and the synthesis of nucleic acids. The characteristic symptom of Mg deficiency is interveinal yellowing which starts from the main vein and develops to the edges. It is usually seen in the lower leaves, due to the ease of translocation to regions of active growth.

3.13.6 Sulphur (S)

The symptoms of S deficiency are seen at the beginning of plant growth. The older leaves become chlorotic and, at the same time, become rigid. Necrosis appears at the apex accompanied by curling of the affected tips and torn edges. Sulphate is readily translocated within the plant.

3.13.7 Boron (B)

The points of active development above ground and in the roots cease to elongate when boron is deficient, and, if the deficiency persists, become disorganized, lose their normal color and die. With death of the buds and the youngest leaves, the adjacent ones become leathery. In general, plants deficient in boron over-produce shoots, with a duplication of the symptoms on the new shoots.

3.13.8 Copper (Cu)

Copper deficiency results in a darkening of the green areas of the leaves. The young leaves are longer and curled down, as though they lack water. Growth seems to be unaffected, at least in the first months of the plant's life.

3.13.9 Iron (Fe)

Cashew tree growth is seriously compromised in the absence of iron. In just one month, the visual symptoms of it deficiency appear, characterized by severe chlorosis in young leaves. With increasing deficiency, the leaves become translucent, remaining light green only in the oldest leaves.

3.13.10 Manganese (Mn)

In the cashew, Mn deficiency symptoms appear initially in the youngest leaves, characterized by pale green colouring, developing later into greenish-yellow and, in some leaves, the edges become brown. Plants deficient in manganese have a small number of leaves and growth slows down although there is a great development of lateral branches. The occurrence of large clusters of small leaves in the shape of a rosette is common followed by the leaves drying, and falling prematurely.

3.13.11 Zinc (Zn)

In the absence of zinc, the plants have short internodes and few lateral branches. In deficient plants, the youngest leaves appear small, elongated, and with a colour varying from green to pale green, but the veins remain green. The lower mature leaves develop normally.

3.14 Irrigation

Cashew is cultivated mainly as a rainfed crop, and areas in which cashew is planted are usually devoid of surface water sources. Although the cashew may grow and produce in regions with an annual precipitation above 600 mm and with a drought of 4 to 5 months, irrigation allows maximum productivity, increasing the harvest period and improving the quality of the peduncle and the nut. Studies in Brazil and other countries have shown that irrigation could increase productivity by up to 300%, depending on the region.

While irrigating cashew plants, care should be taken to see that continuous irrigation is avoided, otherwise trees may remain in vegetative phase for a longer period. Response to irrigation varied among cashew genotypes. Dwarf clones did not respond well to irrigation (Oliveira *et al.*, 2002).

3.14 .1 Irrigation methods

It is recommended to use micro-irrigation (spray or drip), because it has the following advantages over other methods of irrigation: decreased incidence of leaf sickness and weeds, water saving by decreasing losses by evaporation and greater efficiency of water use. Micro-irrigation can also be adapted to different soil and topographies; there is a saving in labour costs and efficient application of fertilizers via irrigation water (fertigation).

Where spraying is used it is recommended to have one jet per plant, with a nominal flow of 30 to 70 L/h and wetting diameter of 3.5 to 5.0 m. In dripping, a minimum of four drippers per plant ought to be used per adult plant in clayey soils, and up to eight drippers per plant in sandy soils.

To choose between spray and dripping as a system of irrigation, the water availability (quantity and quality) should be considered. In dripping there is a greater savings in water and energy, because the loss of water by evaporation from the soil surface is less and the system operates at a lower pressure. On the other hand, the risk of emitter blocking is greater than with spray irrigation, thus better filtering, especially when surface water with a lot of organic matter is used. Dripping also offers the advantage of not wetting the fruits that fall onto the ground, allowing less frequent collecting where the primary product required is the nut. Irrigation at 200 litres of water tree^{-1} once in every 15 days after flowering during summer months increased cashew yield. Irrigating cashew with 60-80 litres of water tree^{-1} every four days through drip after flowering until fruit-set and development, in combination with application of 750:187.5: 187.5g NPK tree^{-1} led to significantly higher yields (Yadukumar *et al.*, 2009).

3.14.2 Water requirements

Schaper et al. (1996), reported that the plant could be irrigated only between flowering and harvest without decreasing yield compared to irrigating during the entire drought period. This saves much water. Ghosh(1995) reported that trial included 30 litres of watering at intervals of 10,20,30,40,50 and 60 days during the period from 31st January to 31st March. Watering at 50 days interval gave the highest fruit set and its retention which consequently resulted 395.6 per cent (Av. Of 2 years) more yield as compared to control. Shelling percentage was improved to some extent due to watering and application of water at 50 days interval showed better results. The water needs of the plant vary with climate, the plant's foliar area, the growth phase of the plantation and with the irrigation method used.

During periods of high evapo-transpiration, 5 litres of water/ day are recommended for each square meter of soil surface shaded by the plant crown or area wet by the emitters (Table.10). The frequency of irrigation depends on the water retention capacity of the soil and should vary between two and four days, for sandy and clayey soils, respectively.

With drip irrigation, the volumes of water recommended in Table.10 may be reduced by about 15%. The number of drippers per plant should increase gradually, according to the age and stature of the plant, from one drip dripper during the first year to up to four, six or eight per adult plant in clayey, medium textured and sandy soils, respectively.

Table 10. Average values of crown projection areas, percentage of soil covered by the plant and volume of water to be applied in irrigation as a function of plant age.

Year of crop	Crown projection area (m^2)	Soil covering (%(1))	Volume of water L/plant/d(2)
1st	1	2	5
2nd	5	10	25
3rd	15	30	70
4th	25	50	120
5th +	30	60	145

1. Assuming the spacing between plants to 7 x 7 m.
2. If the area wetted by the nozzle is greater than the crown projection, the volume of water to be applied should be chosen as a function of the wetted area.
Source: Miranda, F.R. de., 2005; unpublished data.

Soil humidity or water tension monitoring is recommended, in order to ensure that the volumes of water applied and the frequency of irrigation best serve the needs of the plant. Tensiometers can be used to monitor soil water content. For each homogenous area, in terms of soil and cultural phase, tensiometers should be put in at three different locations where there is the greatest concentration of roots. This allows identification of sensors with readings well above or below the average and when this occurs it is necessary to determine if the problem is in the sensor or in the irrigation system (blocked emitters, leaks in the supply lines, etc.).

For the cashew, tensiometers should be installed at two depths in each location of monitoring: the first at 20 cm and the second at 50 cm deep. The distance of the sensors in relation to the tree trunk varies from 30 cm in the first year of crop up to 1.6 m for adult plants. When drip irrigation is used, the tensiometers should be installed a lateral distance of 20 cm from the dripper. The readings from the tensiometers should be performed in the morning, preferentially. For cashews planted in sandy soil, the soil water tension between irrigations should vary between 8 and 25 centibars. For clayey soils, the ideal range is between 30 and 50 centibars. Lower readings in which the minimum values cited indicate that irrigation is excessive. Reading higher than the ideal range indicates that the soil is drier than desirable and the quantity of water ought to be increased and/or the irrigation interval reduced.

3.15 Fertigation

The application of fertilizers through the irrigation water (fertigation) has the advantages of increasing the efficiency of the fertilizers and reducing the costs of labour and machinery for its application. Fertigation allows the application of nutrients with greater frequency, without increasing the cost of the application, minimizing losses by volatilization and leaching and optimizing nutrient absorption by the roots. The nutrients most frequently applied in fertigation are those with greater mobility in the soil, like N and K (Oliveira *et al.*, 2002).

To apply nutrients by fertigation, tanks of the solution, where the fertilizers are pre-diluted in water and an injecting device are necessary. The types of injectors most utilized in fertigation are injector pumps, venturi and differential pressure tanks. There are many advantages to fertigation: a) uniform application of nutrients; b) application of nutrients according to the needs of the plant and the rate of uptake; c) greater efficiency of nutrient use due to its mobility in the wetted zone of the soil where the root system is concentrated; d) savings on labour and agricultural equipment; e) reduction in soil compaction from the use of heavy equipment; f) ability to apply nutrients more frequently thus reducing nutrient losses (Santos *et al.*, 1997). Fertigation needs to be carefully managed to avoid soil acidification and salination in the root zone. To avoid blocking the emitters the fertilizers used should be fully soluble in water and should not form precipitates, especially calcium and iron phosphates.

Water soluble fertilizers like diammonium phosphate, urea and muriate of potash are used for fertigation through drip lines from December to March and application of 2 kg castor cake to soil during August. Fertigation is done once in a week from December to March. With fertigation quantity of nutrients (through fertilizers and organic manure) to be applied can be reduced to half of the quantity of recommended nutrients. Yield can be increased from 1 ton per ha in case of high density planting without irrigation to 2 tons per ha with fertigation of $1/4^{th}$ of the recommended fertilizer nutrients and remaining $1/4^{th}$ applied to soil in the form of castor cake. It was reported that fertigation saved 50% fertilizer requirement and doubled cashew yield (Yadukumar and Mandal, 1994; Mishra *et al.*, 2008).

Kumar et al.(1998) revealed that among the fertigation treatments, split application of water soluble NPK fertilizers at 100 and 80 level recorded maximum nut yield/plant during 2nd year of crop itself (3.72 and 3.82 kg respectively) followed by normal NPK fertilizers (3.09 and 3.03 kg respectively as against nut yield of 2.44 and 1.77kg/plant in conventional method of fertilizers with and without drip system of irrigation respectively.

3.16 Soil and water conservation techniques

In the western and eastern coastal areas, cashew is grown mostly on sloping land. A considerable amount of nutrient leaching and soil erosion are common in such situations. Growers have been advised to construct terraces and contour pits to conserve runoff water. Studies on the extent of root distribution revealed that 90 percent of the root system was confined to a radius of 2m and a depth of 1 m. In order to achieve maximum utilization of applied nutrients, fertilizer practices were confined to this part of the root zone. Before the advent of the south-west monsoon in May-June, basins of about 1.5m radius are prepared for 2 years-old trees and subsequently widened to 2m in the third year. Terraces are made by removing the soil from the elevated portion above the tree trunks to create basins of 1.5-2m. Contour drains are also constructed to collect rain water above the tree-line and prevent soil wash from the slopes.

Yadukumar (2003) reported that such soil and water conservation measures such as, i). Terrace with catch pit (2m length, 0.3m width and 0.45m depth) on upper side of slope dug within three years after planting, with mulch at the base of plant, ii). Coconut husk burial in trenches per plant (1m width, 0.5m depth and 3, 5m length), opened between two rows of the cashew plants, for undulated land in the west coast, and iii). Opening circular trenches (0.3m width, 0.5m depth at 2m away from trunk) within 3 years and burial of coconut husk or leaf litter covered with thick layer of soil or simply ploughing the land in the inter-row space before receipt of summer showers in level areas in east coast, were found to provide moisture required during peak flowering and fruiting season thereby increasing about 25-30 nut yield and hence, recommended for commercial adoption. Similarly, Ghosh (1999) observed that retention of soil moisture in the rooting zone during the critical period of nut setting and maturity stage, by *in-situ* moisture measures in cashew orchards would greatly increase the nut yield. Opening of circular trench around the tree at 4 feet distance with 9″ width and 9″ depth and mulching it with dry leaves helped to retain sufficient sub-surface soil moisture during flowering and fruiting, thereby increasing the yield by 146.7 per cent over control, under Jharagram conditions in West Bengal.

Under medium to steep slopes, 'terrace with crescent bund' and application of 20 kg poultry manure in cashew is beneficial for improving soil moisture content, yield and also reduced runoff and soil loss (Yadukumar and Rejani, 2006). Reduction in peak runoff and increase in recession time and groundwater recharge due to soil and water conservation practices have been reported by several workers (Deshmukh *et al.*, 1992). Black polythene mulch was helpful in conserving soil moisture (Nawale *et al.*, 1985).

A gradual decline in morphological traits was noticed with increase in soil moisture stress in cashew (Latha and Abdul Salam, 2003). Leaf area and total chlorophyll contents were observed to decrease with increases moisture stress and resulted in low total dry matter production. Relative water content, though declined, was not conspicuous at lower level of stress However, root: shoot ratio remained unaffected upto 90 days with depletion of available water, indicating the capacity to withstand varied degree of soil moisture stress.

Using coconut coir pith as soil mulch in cashew plantations resulted in 14.15% higher water retention and suppression of weeds upto 73.52% (Kumar *et al.,* 1989). Badhe and Mayar (2004) reported that trapezoidal shaped staggered trenches (230 ha^{-1}) of 4.5 m length, 0.60 m top-width, 0.30 m bottom-width and 0.30 m depth were effective in reducing runoff and for conservation of soil and nutrients. Mane *et al.* (2009) demonstrated that continuous contour trench (0.5 m x 0.6 m) was the best soil-conservation practice for cashew on hands with 7 to 8% slope.

Yadukumar and Balasimha, (2006) reported the physiological benefit of irrigation and fertiliser management in cashew for better yield performance. It was observed that the adequate soil moisture within available range at deeper depth during December, even in non-irrigated control plot, facilitated the trees for normal photosynthesis (10.14 - 12.01 ìmol m^{-2} s^{-1}) and did not affect microclimate. Later, the transpiration (9.93-10.96 mmol H_2O m^{-2} s^{-1}), photosynthesis (14.50 – 17.10 μ mol m^{-2} s^{-1}) and stomatal conductance (0.443-0.671 mol m^{-2} s^{-1}) were observed to be significantly higher in trees which received irrigation and fertilizer application than that of control trees during March. This was attributed to higher air/leaf temperature, lower relative humidity and moisture stress below the lower level of available moisture in un irrigated control plot. These results indicated that despite inherent ability to withstand stress conditions, as has been observed earlier, cashew responds to favourable conditions, if provided, resulting in higher physiological output.

3.17 Weed management

Weeds are one of the most important pests of cashew plantations. The main task after establishment of cashew plantation is to control excessive weed growth which might otherwise severely deplete soil moisture, compete for nutrients, pose a problem in harvesting of nuts and apples. Considerable expenditure is required to keep the plantations weed free. Depending upon the type of weeds and intensity of weed growth, weeding is to be done manually, mechanically or chemically. The weed growth is seen maximum after the monsoon (October-November) and harvesting becomes a problem if the weeds are not removed.

3.17.1 Weed Flora

The dominating weed flora in cashew plantations is mostly dicots. Monocots are also present in certain areas. Among dicots, *Eupatorium odoratum* (Compositae) is most common. Other dicots include *Tridex procumbens* (Compositae), *Mimosa pudica* (Mimosae), *Lantana indica* (Verbenaceae), etc., Among monocots, *Cyperus compressus* (Cyperaceae), *Setaria glauca, Avena sativa, Cynodon dactylon* (Gramineae) and *Silax zeylanica* (Liliaceae), etc., are found in the plantation.

3.17.2 Management

Weeding around the young trees within a circle of about 1.5 m diameter by hoeing is necessary during 2-3 years after planting. It is beneficial to weed the young trees twice a year. The weeding cycles are generally confined to the pre-monsoon and post-monsoon periods to coincide with the fertilizer application. The first round of weeding may be done, before heavy rains start, by uprooting the weeds within the radius of 2m from the main stem. The remaining weeds in the interspaces are slashed to ground level. The second round of weeding may be done during November-December to facilitate spraying, harvesting and picking of nuts. In grownup orchards, careful weeding by clearing the area within 2 m radius of the trunk and slashing the remainder is essential until the trees shade out most of the weeds.

3.17.3 Manual weeding

Manual method like hand pulling, sickling and mulching are extensively used wherever labour is available. Hand pulling is the most efficient in annual and biennial weeds which are propagated by seeds. Sickling is done to remove the tall growth of perennial weeds and to cut to the ground level for a follow up application of foliage applied herbicides on the regrowth.

3.17.4 Mechanical weeding

Mechanical hoeing may be carried over the whole-planted strip to a width of about 4 m, leaving 2 m in both sides, to avoid root injury. A tractor-mounted weed slasher can also be used to control weeds effectively and economically. Tillage practices like hoeing, ploughing and mowing are done to clear off weeds, especially in newly established orchards where sufficient spaces may be available for inter cultivation. Mulches like straw hay, farm wastes are used to cover the soil. They produce heat which smothers or kills the weeds. Kumar *et al.* (1989) demonstrated that mulching with coconut coir-pith to a depth of 7.5 cm resulted in retention of about 14% more moisture and suppressed weed growth. Heavy mulching of young cashew trees at Kununurra was partly effective in controlling weed growth, but was laborious to apply (Millington, 1992).

Adeyemi (1989) described using intercropping with food crops (maize, cassava, cowpea, plantain bananas) to reduce weed incidence in cashew crops in Nigeria during the early years after transplanting the young trees to the plantation. Similar techniques using forage sorghum as the inter-crop were reported by Heading (1992) to be successful in reducing weed growth in young trees during establishment at Kununurra.

3.17.5 Chemical weed management

Alternatively, weedicides may also be applied after slashing, well in advance of the rainy season if the under growth is too dense. Chemical weed control is costly and thus reserved for exceptional cases, especially in areas where the labour wages are high. Effective chemical weed control methods are available for cashew plantations. Application of Paraquat @ 0.4 kg per ha twice at bi-monthly intervals starting from July will effectively control all types of weeds. Application of Glyphosate @ 0.8 kg per ha, in June–July can also control the weeds effectively. The recommendation is to apply Agrodar-96 (2,4-D) at the rate of 4 ml/litre of water followed by Grammoxone at the rate of 5ml/litre of water. Approximately 400 litres of spray is required to cover one ha. The spraying is repeated in the post monsoon season if the weed load is heavy.

3.18 Training and pruning

Pruning an essential horticultural technique for the improvement of crop yield in several horticultural crops (Mika, 1986) was not well understood earlier in cashew. Later studies have proved that pruning provides a definite form and shape to the tree. To achieve better fruit set, young cashew trees are often trained to provide a better tree form.

Majority of the cashew plantations were established without proper training and pruning or canopy management. Such plantations become unmanageable due to erratic branching and uneven spreading of canopy and finally leading into cashew forests. The lower branches die due to shading effect resulting in less number of flowering and fruiting branches per unit canopy area. In order to rejuvenate trees of with low yield heading back by severe pruning can be adopted in order to improve the yield performance provided they are potential yielders. Heading back is to be done after the annual harvest of the crop and the trees get ready with a new canopy for flowering and fruiting in the ensuing flowering season.

Training of young trees during the first three years is essential to develop uniform canopies. Training in the juvenile phase comprises of removing basal branches and water shoots. The plants are trained to a single stem and branches are allowed to grow about 0.75-1m from ground level. Deformed branches are also removed during the first few years. Since cashew trees tend to spread their canopies and lodge easily, proper staking is also essential. Orchard operations such as terracing, weeding, fertilizer application, nut collection and stem/root borer infestation control can be easily achieved if trees are properly trained.

Depending on the variety the flowering on the severely pruned trees takes place either in the same year or next year. Pruning should be carried out in August-September at least once in three years, when unwanted growth is removed to provide adequate sunlight into the canopy. However, precaution needs to be taken to monitor each plant for the attack of cashew stem and root borer on the severely pruned trees. So, pruned wounds should be swabbed with 10% Bordeaux paste to avoid fungal disease and pest attack (Nalini, 1997). Since fruiting is only encouraged from the third year, de-blossoming has to be carried out as flower clusters appear during the juvenile phase.

Mohan (1983) had also noticed a higher number of bisexual flowers in the pruned shoots of cashew. Khan *et al.* (1987) reported that pruning of the dead wood alone helps in improvement of 30-40% nut yield in cashew. While, Mohan and Singh (1988) reported that the nut yield can be doubled by leader shoot pruning. Indiscriminate pruning adversely affects fruiting (Sathpathy, 1988). More number of lateral productions in the pruned shoots was also observed in cashew (Mohan and Singh, 1988). Pruning treatment increased the number of laterals/leader but did not affect duration of flowering and harvest (Mohan and Rao, 1995). Leader shoots pruning at least once every 2-3 years helps boost nut yield (Nayak, 1996).

Aneesarani *et al.* (2011) reported that tertiary shoot pruning along with foliar spray increased yield by about 30 per cent in cashew. Pruning of cashew tree influenced the production of flowering laterals, number of bisexual flowers per panicle and the number of fruits per panicle (Ghosh and Chattopadhyay, 1999). Tertiary pruning recorded higher number of flowering clusters per square meter in cashew var. VRI 3 (Aneesarani *et al.*, 2011).

3.19 Top-working

Mostly old cashew plantations are of seedling origin and have become senile. Further, cashew plantations raised from grafts of varieties not recommended for the region also give low yield. Such trees can be rejuvenated by grafting with scion sticks of improved and high yielding varieties on new shoots arising on beheaded stumps. Trees of 20 to 25 years old are beheaded. Top worked trees start flowering and fruiting in the very next year after grafting with new vigour. May-June is the right time for beheading and July-August is for grafting. However, precaution needs to be taken to monitor each plant for the attack of cashew stem and root borer on the top worked trees. So, after beheading stumps are protected with Blitox and Sevin 50 WP and shade should be provided. Profuse sprouting normally results in but only 10 to 15 healthy shoots and properly spaced on the stumps are alone retained. These shoots are grafted at softwood stage(cleft grafting) when they are about 40 to 50 days old. 7-8 successful grafts may be encouraged to grow and the sprouts should be periodically removed. Top worked trees grow vigorously due to the well established root system and they start yielding about 4 kg per tree from the second year of rejuvenation and the yield gradually increases to stabilize at 8 kg from the fourth year of top working. Precocity and increased productivity are advantages of top-worked trees. This technique can also be utilized for establishment of scion banks, since this is a quick and viable method for mass production of scion materials.

The technique of rejuvenation of existing, unthrifty cashew plantations by top-working boosts cashew production 3-4 fold in a short span. This technology can also be adopted for mass production of scions and production can be expected to be almost five times that by conventional methods (Khan *et al.*, 1988). Top-working trials in red and laterite zone of West Bengal showed that beheading of trees (Fig. 10) at 1.0 m height was ideal from the point of view of sprouting, growth of sprouted shoots and graft success. The month of October was most suitable for beheading and February for grafting, irrespective of age of the tree (Ghosh, 1991). Whereas, under Ullal conditions of Karnataka, December to February was found to be suitable for beheading and February to April for grafting (Khan *et al.*, 1985). Uneconomical cashew trees top-worked with high-yielding Ullal-1 variety resulted in a four-fold increase in nut yield per tree within five years (Kumar, 1990). Wherever populations of cashew stem and root borer (CSRB) are maximum, top-working technology may not be suitable, unless regular follow-up action is taken to manage the incidence of CSRB (Swamy, 1995).

Fig.10. Rejuvenation upon beheading

Unproductive 20 years old VRI-1 cashew trees top worked with VRI-3 scions at RRS, Vriddhachalam has resulted in harvest of the potential yield of VRI-3 *ie.,* 8 kg/tree. Management control of CSRB has also been successful in these trees with regular monitoring, swabbing with pesticides at the base of the trunk during egg lay period.

3.20 Bio fertilizers

Application of *Azospirillum, Azotobacter* and VAM @ 10 g/poly bag at the time of sowing cashew nuts increased germination percentage of nuts and plant growth, and reduced the incidence of fungal diseases in the nursery (Kumar *et al.*, 1998; Ramesh *et al.*, 1999; Aneesarani and Jeeva, 2010). Inoculation of *Azotobacter* resulted in higher root growth (Oblisami *et al.*, 1985) and yield (Singh, 1997) in cashew. Among VAM, *Acaulospora laevis* and *Gigaspora mosseae* were better symbionts for inoculating cashew with (Lakshmipathy, 2000). Anantha Krishnan *et al.* (2004) reported that among VAM, *Glomus fasciculatum* was superior in terms of increased shoot-length, internode number, number of leaves, stem diameter, root length and root number, under nursery conditions. Sivaprasad *et al.* (1992) also reported that *Glomus fasciculatum* to be more effective at enhancing growth and P uptake of cashew plants. VAM (25g/bag) was helpful for better graft-uptake at the time of grafting (Sridhar *et al.*, 1990).

Roots and Rhizosphere soil samples were collected from NRCC cashew Plantation, Puttur and Screened for the presence of vesicular and arbuscular mycorrhiza (VAM). Pure cultures of the dominant species were made using sorghum as host plant. Roots collected from field grown trees as well as isolated seedlings were found naturally infected with VAM as revealed by the presence of vesicles, arbuscules and mycelium. Four species of VAM isolated from cashew soils were identified as *Glomus macrocarpum, G. australe, G. multicaule* and *Scutellospora nigra.* Among the 4 species, *G. macrocarpum and S. nigra* were widely distributed and found associated with the roots of cashew trees as well as isolated seedlings. Hence, these two species could be used in the acclimatization of micropropagated plants (Anil *et al.*, 2000).

Azotobacter and *Azospirillum* inoculation enhanced the germination, growth and development of cashew seedlings in the nursery. In general, *Azospirillum* inoculated plants were more efficient than *Azotobacter* inoculated cashew plants. The same trend was observed in the seedlings and grafts planted in the field. An increase of 24.03% in yield due to combined inoculation of *Azotobacter* and *Azospirillum* could be observed in twelve years old cashew trees.

Obilisami *et al.*(2011) reported that a pot-culture trial was undertaken to study the effect of *Azotobacter* inoculant either alone or in combination with the plant growth regulators like GA and IAA on cashew. The combination of *Azotobacter* incoulant alone caused better growth of root system when compared to that due to growth regulators alone. Similarly, better shoot length was observed in treatment receiving *Azotobacter* inoculant alone. The population of *Azotobacter* in the rhizosphere was more in treatments receiving *Azotobacter* alone and *Azotobacter* plus GA indicating the establishment of *Azotobacter* in the rhizosphere upto 75 days of plant growth.

3.21 Plant growth regulators

The problem of declining tree yield has led to an investigation into the effectiveness of foliar application of exogenous hormones to improve flowering, fruit set, and fruit retention in cashew. Foliar spray of growth regulators Planofix, Nutron, IAA, IBA, NAA, 2,4-D and Ethrel were effective in increasing total number of flowers, hermaphrodite flowers, sex ratio, fruit and yield per panicle and also improved physico-chemical composition of apples and nuts (Ghosh, 1988; Singh *et al.,* 1992 and Kumar *et al.,* 1994).

The extracts of cashew seeds when chromatographically assayed, showed to contain natural auxin IAA with an Rf of 0.30 and is very much similar to synthetic IAA. The effect of plant growth regulators viz. GA, IBA, NAA and CIPA at 100 to 500 ppm for seed treatment, 25 to 100 ppm spray treatment and 0.1 - 0.4 per cent in lanolin paste on the seeds and seedlings was studied. Invariably all the growth regulators except CIPA stimulated high percentage of seed germination. CIPA at all concentrations inhibited the germination and root development. GA induced stem elongation in seedlings was due to increase in the cell number and cell elongation as observed in the cortex and pith regions of the stem sections of the second internode. The fresh and dry weights of shoots and roots showed increase with GA treatment.

3.22 Cover cropping

Popular leguminous cover crops for cashew plantations are *Peuraria javanica, Calapagonium muconoides* and *Centrosema pubescens* which improve the fertility and moisture balance and help conserve orchard soils. Cover crop seeds are generally sown with the advent of the monsoons at a seed rate of about 7 kg/ha. Seeds should be soaked in water for six hours before sowing and are sown in 30 x 30 cm beds that are prepared in interspaces of cashew. These cover crops will also help in checking soil erosion. As cashew responds to nitrogen, cultivation of leguminous cover crops will also enrich the soils with nitrogen, which will beneficial to growth of cashew. However, it should be ensured that basins of cashew trees are cleared before harvesting season, so that growth of cover crops does not interfere with collection of nuts. On degraded steep lands, cover crops are usually established on seed beds between tree rows.

3.23 Inter-cropping

Inter-cropping has become popular with the systematic establishment of large -scale orchards. It is practised in the first few years when there is sufficient space between crop rows with the main objective of deriving some income until the cashew starts giving economic returns. In Andhra Pradesh, popular inter-crops are horsegram, cowpea, groundnut etc. *Casuarina* is also a tree inter-crop planted at a spacing of 1.5m x 1.5m in cashew orchards in Andhra Pradesh and Odisha. In the west Godavari district of Andhra Pradesh, cashew is grown in combination with casuarina and coconut.

In Kerala and coastal Karnataka, pineapple is grown as a biennial crop in the initial 4-5 years and farmers find it far more profitable than crops such as redgram and cassava. Adopting a two-row system of planting in contour trenches, yields of 15-20 tons/ha have been achieved from this inter-crop. The pineapple inter-crop also indirectly benefits the main cashew crop as contour trenches help to conserve water and nutrients. When hedge-row planting at a spacing of 5m x 10m is used, inter-crops of Acacia, casuarina, etc. are taken without any ill-effects on the main crop. These trees are planted about 3.5 m away from the cashew in two-row plots spaced 1m x 1m apart in the center of crop avenues.

Intercrops like cowpea, French bean, cluster bean, rice bean, red bean, mung bean, soyabean and groundnut could be grown along with cashew (Gupta, 1999). Maize and groundnut can be grown successfully as inter-crops in newly-planted and two years-old cashew orchards (Abeysinghe *et al.,* 2003). Inter-cropping cashew with pineapple, turmeric or elephant foot yam under normal density planting system during the first five years increased net benefit from cashew gardens (Yadukumar *et al,* 2003). Weed-suppression was best in plots carrying cashew/cassava and cashew/plantation/cassava mixtures with 50-60% reduction in frequency of weeding per annum (Adeyemi, 1998).

Pineapple can be grown as intercrop in cashew garden profitably for the first 7 years. Both main and intercrop can be planted simultaneously. Normally pineapple can be grown as inter-crop in the interspace available between two rows of cashew plants. Here spacing maintained for cashew is 8m x 8m (156 trees/ha) or 7.5 m x 7.5 m (175 trees/ha) or 10 m x 5 m (200 trees/ha). The net return from cashew orchard in the initial years of plantation can be enhanced through growing pineapple as inter-crop. Other advantages of growing pineapple in trenches dugout across the slope between two rows of cashew are better soil and water conservation which will have indirect benefit over main crop (cashew) resulting in 1.5 times increase in yield as compared to plot where cashew alone is grown and low weed growth. Pineapple can be grown as intercrop under irrigated as well as unirrigated conditions in west coast region where rainfall received is from 3000 to 3500 mm distributed over a period of five months. Other crops such as tapioca, turmeric, ginger, cucurbits, *Colocasia,* and Elephant Foot Yam have been found to be suitable intercrops in West Coast region while *Colocasia,* brinjal, groundnut and black gram were found to be good intercrops in East Coast region.

Ohler (1979) recommended the use of early bearing low growing food inter-crops and suggested that tall inter-crops like certain varieties of sorghum (*Sorghum bicolor* (L.) Moench), and millet should not be grown with cashew as they cast too much shade and affect the growth of the latter. However, in Nigeria crops that vary in height like maize, cassava (*Manihot esculenta* Crantz), bambara groundnuts (*Vigna subterranea* (L.)Verdc.), cowpea (*Vigna unguiculata* (L.) Walp), melon (*Citrullus lanatus* (Thunb.) Mansf.) and plantain (*Musa* spp.) have been successfully grown along with cashew.

Opoku-Ameyaw *et al.* (2011) reported that Intercropping significantly improved the girth of cashew seedlings in the cashew+ maize combination. Height of cashew seedlings was significantly higher in the cashew + maize and cashew + sorghum/groundnut system. Intercropping did not substantially alter early yield of cashew nuts. Economic analysis of the packages, however, indicates that maize and yam inter-cropping during the establishment phase was profitable. It is, therefore, advisable to intercrop cashew with yams and maize in northern Ghana during the early period of establishment to generate additional returns to the growers.

Kumar *et al.*(2012) Field crops such as groundnut (*Arachis hypogeae*), blackgram (*Phaseolus mungo*), greengram (*Vigna radiata*), vegetables such as cucumber (*Cucumis sativus*), bottlegourd (*Lagenariasiceraria*), tuber crops and fruit crops such as pineapple (*Ananas comosus*), spices such as turmeric (*Curcuma longa*), ginger (*Zingiber officinale*) and pepper (*Piper nigrum*) have been found to be suitable and profitable intercrops in cashew plantations, increasing the total return per land unit during the early stage of establishing a cashew plantation.

Aneesarani *et al.* (2011) experiments on inter-cropping medicinal plants along with young cashew plantations revealed that among the inter-crops experimented namely *Ocimum sanctum, Aloe vera, Phyllanthus niruri and Catharanthus roseus,* the two inter-crops *Ocimum sanctum* and *Aloe vera* recorded higher BCR values of 3.0 and 2.9 respectively performing better as inter-crops in cashew. *Ocimum* showed sustained performance for three years and *Aloe vera* for two consecutive years. These results showed that *Ocimum* and *Aloe vera* could be promoted as profitable intercrops during the pre-bearing age of cashew.

Keisarlourdusamy *et.al.* (2016) reported that direct sown vegetable crops such as bhendi (*Abelmoschus moschatus*), cluster bean (*Cyamopsis tetragonoloba*) and greens (*Amaranthus* sp.) can be profitably grown as inter-crops with benefit cost ratio more than 2.0 in early establishing periods of cashew plantations.

3.24 Organic Cashew

Organic cashew cultivation is still in its infancy. But it is fast catching up in parts of south India due to the 20% to 25% price advantage over conventional method of growing the cash crop. The demand in the US, European Union and Japan, is also expediting the process of switching over to organic cashew cultivation. Fuelled by the need for pesticide-free products, the demand in these countries far outstrips the supply with certified organic products commanding a premium price of 25% to 30% more than the normal yield. Globally, of the 700 tonnes organic cashew trade, Brazil supplies 500 tonnes with India contributing the rest. Goa and parts of Kerala have been producing organic cashew with success, the study further said. With most cashew producing countries, gearing up to implement a programme for organic cashew cultivation, large-scale agri-export zones are likely to play a key role in it.

Experiments conducted under All India Coordinated Research Programme on cashew at Regional Research Station Vriddhachalam revealed that pure organic farming of cashew in field condition though results in increased cost of cultivation, the yield is comparable to integrated nutrient managed fields. The quality of cashewnuts obtained also superior in organically managed plantations Keisar-lourdusamy *et al.* (2016).

In major cashew growing areas of the country, natural farming is followed. It is estimated that not more than 20 percent of cashew holdings in India use chemical fertilizers/pesticides (Sivaraman and Hubballi, 2002). This facilitates adoption of organic farming practices over a period of time without additional efforts. Thus, there is a vast potential of bringing these areas under organic farming practices to exploit the available marketing avenues. The Department of Horticulture in Tamil Nadu is promoting the use of organic inputs like bio fertilizers, neem cake, vermi compost, etc. The sample farmers were found to use the organic inputs initially but later used the chemical fertilizers; hence they were considered as partially organic farms.

The most important producing countries for organic cashew nuts are Brazil, Madagascar and Sri Lanka. In Mexico and Honduras, organic cashew nut cultivation is still relatively unknown.

Cashew trees place no special demand on soil fertility. When, as usual in organic cultivation systems, cashew trees are cultivated with bottom crops in agroforestry systems, then the resulting organic material provided from pruning and green manure plants is sufficient. Additional organic fertilizer is then usually not necessary. Conventional examinations have resulted in the following nutrient-depletion values for trees with a yield of 24 kg Nuts per Tree: 2.85 kg N, 0.75 kg P_2O_5 and 1.27 kg K_2O (Ohler, 1979).

3.24.1 Organic recycling

Cashew plantations have a vast potential for organic biomass for recycling. Availability of cashew leaf-litter from plantations of different age groups (10 to 40 years) ranged from 1.38 to 5.20 t ha^{-1} (Guruprasad *et al.*, 2009). It is the main source of maintenance of soil productivity in cashew plantation but decomposition is incomplete and slow. An estimated production of 8 lakhs tonnes of cashew apple is also discarded every year. About 5.5 t ha^{-1} available cashew- biomass waste can be converted into 3.5 tonnes of compost or vermicompost and helps to meet nutrient requirement of cashew plants 50% (Yadukumar and Nandan, 2005; Yadukumar, 2007). The cashew leaf waste compost contains 1.59% N, 0.53% P, 0.33% K, 0.94% Ca, 0.56% Mg, 4.02% ppm Cu, 74.5% ppm Mn and 86 ppm Fe. Nutrient composition of compost prepared from cashew apple and litter enriched with sawdust, cowdung slurry and poultry manure in 3:3:2.2 was analysed as 1.67% N, 0.44% P, and 0.58% K. The pH of prepared compost was 8.9, which will be a good ameliorant for acidic soils (Mini *et al.*, 2005).

Kumar *et al.* (2012) reported that cashew is a deciduous tree that provides approximately 5 tonnes of cashew biomass residues (leaves, twigs, flowers and apples) per hectare in a well-established cashew orchard. Use of earthworms for the production of vermicompost from cashew biomass is a low cost technology with great potential; 3.5 tonnes of vermicompost can be produced per ha of adult cashew orchard per year.

3.25 Dwarf cashew

Extensive cultivation of high yielding varieties/hybrids under high density planting system is the best option to double the cashew production. On the other side, the existing cashew varieties are mostly of spreading types and necessarily needs annual maintenance pruning, which is time and labour consuming. Though dwarf cashew types are reported from Brazil, such types are not located in India. Introduction of dwarf types from Brazil and their utilization in breeding programmes is a priority area of research. Other method such as mutation and chemical induction of dwarfism are reported. Effect of gamma irradiation on cashew seeds showed that seedlings produced from nuts irradiated at doses of 40-60 Gy were dwarf. Foliar application of cultar 1000 mg l^{-1} along with KNO3 1% and carbaryl 0.1 %@ 5 l per tree to mature lateral shoots of cashew before flushing (September) induces early flowering by one month compared to untreated trees. This spray application caused profuse and synchronized flowering resulting in two-fold increase in nut yield compared to control.

A smaller, more compact tree is more efficient at intercepting the sun's energy (Green *et al.*, 2003). Development of dwarf hybrids or genotypes with compact canopy can suit to the high density planting system to enhance the yield of nuts per unit area. In India, dwarf / compact canopy hybrids are yet to be developed or released (Bhat *et al.*, 2007). Reduced stem elongation, decreased leaf size and dark green foliage in GA-deficient dwarf hybrid poplar mutants were reported by Busov *et al.* (2003). They observed four-fold reduction in height in these mutants. These dwarf varieties allows the plant to put more energy into the useful organs instead of stalks or trunk and reduces toppling damage by wind and rain (Moffat, 2000).

Aneesarani *et al.* (2011) reported that the hybrid HC 6 was dwarf with 3.0 m height, reduced number of primary and secondary branches, number of leaves per current season's shoot (5.3), reduced internodal length (0.25 cm) and number of internodes (5.3). HC 6 also recorded a higher number of stomata (57.2 per mm^2). The highest cumulative yield for three years was recorded in HC 1 (3.5 kg nuts per tree) followed by HC 6 (2.65 kg nuts per tree). HC 1 was semi-tall and compact. Correlation studies revealed that there is strong positive correlation between plant height and length of current season's shoot, number of internodes per shoot, internodal length and number of branches. This study resulted in a hybrid with reduction in height, number and length of internodes, spur formation on shoots, reduced leaf size, number of shoots with moderate yield. However, the genetic nature of dwarfness is to be confirmed by further evaluation through conventional breeding and molecular approaches.

3.26 Harvesting

The harvesting and processing of cashew is very labour intensive. Flowering in cashew starts from third year onwards and reaches full production potential by tenth year and gives economic yield for a further 20 years. After fruit set, nut develops quickly and reaches full size 3-4 weeks after flowering. Pedicel enlarges to produce cashew apple. Fruits mature in 60 days and they are harvested when fully ripe. Harvesting generally involves collecting the nuts once they have dropped to the ground after maturing. Workers scour the area and detach the nut from the fruit. The nuts are generally collected in baskets or sacks. Although activities are labour intensive and time consuming, they are not heavy and women and children can help.

Nuts are separated from apple, dried in sun for 2-3 days and stored temporarily. Flowering commences in November and extends upto February. The peak months of harvest are March-April and the remaining crop comes to harvest in February and May. Yield of nut varies enormously from 1 to 45 kg/tree/year. Though national average of India is 3 kg/tree/year, high- yielding clones yield upto 17 kg. Yield of cashew apple is 8-10 times of nut yield.

The nuts collected should be dried immediately under sun by spreading in a thin layer to reduce the moisture content of the nut. If the surface is of cement concrete, drying for two full days is sufficient. If the surface is of mud, drying for 3 to 4 days is necessary. While drying, the nuts should be raked quite often. Sundrying of cashew nuts is done on specially prepared drying floors, The drying areas are smooth and slightly sloping, so as to allow rainwater to run off. The nuts sare constantly raked in order to ensure that they all receive the same benefit of the sun's rays and therefore, they are dried evenly. Nuts should not be dried for more than four days since they become brittle and break while processing, causing damage to the kernels. Cashewnuts should keep for 12 months or more. Moisture content may cause deterioration of the kernel due to mould or bacterial attack, or enzyme action. The nuts are, hence, dried after harvesting to preserve their quality.

❑❑❑

Chapter 4

Plant Protection

Pest and Disease Management in Cashew

India is the largest producer of Cashewnut (*Anacardium occidentale L.*) earns an attractive export value in the foreign markets as a major cash crop. During the past two decades significant developments have been taken place in standardization of production technologies to improve the yield potential of cashew. In India, the productivity of cashewnut is impaired largely by problems associated with attack of pests and diseases, which is quite apparent from the fact that the production trend is not in pace with the increase in area of cultivation. As many as 70 species of insect pests and 20 diseases are reported to infest cashew in various stages of growth (Pillai *et al.*, 1976). Among the insect pests, cashew stem and root borer, tea mosquito bug, mealy bug, leaf thrips, flower thrips, blossom webber and hairy caterpillars cause substantial yield loss and lower the profit of the growers. Adequate control measures need to be adopted for key pests so as to attain maximum returns.

5.1 Major Pests

5.1.1 Stem and root borer, *Plocaederus ferrugineus* L. (Cerambycidae: Coleoptera)

The cashew tree borer, *Plocaederus ferrugineus* (Fig.11) is a destructive pest throughout the year in all the cashew areas surveyed. (Basu Choudhuri and Misra, 1962). The adult beetle lays its eggs in crevices of loose bark in the trunk. The grub that hatches out bores in the bark and feeds on the sub-epidermal and vascular tissues and tunnels are formed. The affected trees show the symptoms of oozing out of the gum, extrusion of chewed up fibers and excreta, yellowing and shedding of foliage. The borer attacked trees are killed subsequently, accounting an average of 5-12% loss of productive trees every year in different areas. In certain neglected plantations, even up to 26 per cent damage was observed. Apart from *Plocaederus*, two other allied species of stem borers *viz., Plocaederus obesus* and *Batocera rufomaculata* were co-existed in most of the cashew areas (Senguttuvan, 1999).

5.1.1.1 Nature of damage

Stem borer infection could be identified by the oozing of brownish resins, extrusion of frass at the collar region, gradual yellowing and shedding of leaves, drying up of twigs and finally death of the tree occur. Adult beetles lay eggs in the crevices of bark on the trunk. The young grubs that hatch out bore into the bark and feed on the sub-epidermal and vascular tissues and the tissues are tunneled in irregular fashion. As a result of the injury to the bark tissues, gum oozes out which will get hardened subsequently resulting in gummosis. When the vascular tissues are damaged the ascent of plant sap is arrested and the leaves become yellow and shed. In the advanced stages of infestation twigs dry up and the tree dies. Cashew trees of more than two years age are prone to be attacked by this pest. However, the infestation, is severe in the older and neglected plantations. Even though the infestation is noticed throughout the year, the peak period of infestation was observed during summer months up to 15 per cent.

5.1.1.2 Bio ecology

The adult female beetle lays around 60 -90 eggs. The eggs are whitish measuring about 3 mm in length (looking like rice grain) and these eggs are laid into the live tissues in the crevices of the bark at the collar region. The egg period varies from 4-7 days. The eggs hatch out as tiny grubs, which bore into the fresh tissues of the bark, feed on the sapwood tissues and make tunnels in broad and in irregular directions. The tunnels are deepest in the middle and shallow at the sides. The grubs feed inside the tissues for 4-7 months. The grown up grubs are off-white in colour measuring about 7-10 cm in length. The fully fed grubs descends to root zone through tunnels, bores into the heart wood and forms oval shaped chamber with a circular exit hole for the emergence of adult beetles. Adult emergence occurs from November-June coinciding with monsoonal rains. The adult beetle lives for 1-3 months. Pest infestation has been found to be heavier in summer than in the rainy season.

5.1.1.3 Management

Integration of phytosanitation, prophylactic, pest-prophylaxis, biological and chemical methods are essential. These measures adopted as a community effort is essential to bring about effective control.

5.1.1.4 Phytosanitation

Field sanitation including uprooting of borer attacked dead trees and severely infested trees beyond recovery stages before and after monsoon season is important. Trials conducted at Research Centres of All India Co-ordinated Research Project (AICRP) on Cashew have confirmed that inclusion of phytosanitation measures was more effective in reducing tree borer infestation rather than the prophylactic treatments to the trunk and exposed roots (Raviprasad *et al.*, 2009). Removal of borer attacked dead trees and trees with advanced stages of infestation may reduce the spread of borer in subsequent periods. Such measures are more effective in reducing the infestation, rather than prophylactic and chemical treatments.

5.1.1.5 Prophylactic control

Prophylactic swabbing of coal tar + kerosene (1:2) on the trunk (from the base upto a meter height) and exposed roots, once in six months is essential to prevent egg laying by adult beetles (Samiayyan *et al.*, 1991; Mohapatra and Jena, 2008). Treatment of cut wounds with maskers like neem oil 5% is recommended for further reducing the borer attack (Chakaraborti and Chakaraborti, 2009). Making of any wound on the tree should be discouraged, as it is to these wound that the female beetles get attracted for egg laying. Neem oil 3% or coal tar + kerosene (1:2) or chlorpyriphos are good protectant materials for treating cut and wound surface of trees to provide protection against the pest.

5.1.1.6 Post-prophylaxis control

Mechanical control involves examining the trunk for borer infestation and extracting the grubs, if present, by means of sharp metal knife. If infestation is detected in the initial stages, mechanical extraction of the grubs by chiseling out the damaged area of the bark, and thorough swabbing of that portion with 3% neem oil (dissolve 30 ml of neem oil one litre of water) is advisable. Spot drenching of diluted chlorphriphos 20 EC @ 20 ml or monocrotophos 36 WSC

@ 20 ml mixed in equal volume of water per tree was also found to effective. Ensuring the availability of skilled manpower resources for carrying out such jobs in cashew orchards is a matter of critical importance.

Cashew Stem and Root Borer Life stages and damage on cashew trunk

Tea mosquito bug and its infestation symptoms

Fig. 11. Cashew stem and root borer and tea mosquito bug (Pictures courtesy: Dr.V.Ambethgar, Professor of Entomology, TNAU)

5.1.1.7 Biological control

Cashew tree borers have only few fungal agents as natural enemies. The bio-stages of *P.ferrugineus* are susceptible to natural mycoses of *Metarhizium anisopliae, Beauveria bassiana, Aspergillus flavus and Aspergillus niger* under field condition (Ambethgar, 2010), particularly when the climatic conditions such as low temperature and high humid weather prevail. A method has been developed to mass culture the *M. anisopliae* and *B. bassiana* on sorghum grains and rice bran, supplemented with urea as a source of nitrogen, in autoclavable polythene bags and reusable saline bottles. Mass cultured fungal spawn can be applied prophylactically to the soil in the plantation at the rate of 250 g/tree. Mixing the spawn with organics like farm yard manure, neem cake and leaf litters can support the inoculums build up in the field.

5.1.1.8 Chemical control

The success of the curative treatment depends much on the intensity of infestation. The effective control measures consist of mechanical removal of the immature stages (grubs) of the pest during initial stages of infestation. Earlier, spot drenching the soil around the trunk of the tree may also be done with carbaryl 50 WP 0.2% solution (@ 4 g/lit of water) (Senguttuvan, 1999). Such measures are more effective in reducing the infestation, rather than prophylactic and chemical treatments.

5.1.2 Tea mosquito bug, *Helopeltis antonii* Signoret (Miridae: Heteroptera)

The tea mosquito bug (TMB) (Fig.11) is considered to be the most serious pest of cashew in India, which cause serious economic losses to the crop than any other pest. It is estimated that this pest alone is responsible for damage of nearly 25% of shoots, 30% of inflorescence and 15% of tender nuts. It causes more than 30% economic loss by inflorescence blight and immature nut fall. Neem is known to be the main host plant of this pest. The alternate hosts of the pest are guava, cocoa, mahogany, cinchona, cotton, apples, grapes, drumstick, black pepper and jamun. Recently, Maximum recovery was observed in Chlorpyriphos 50 EC (0.2%) treated trees followed by Triazophos 40 Ec (0.2%) treated trees (Jayaprabhavathi *et al.*, 2016).

5.1.2.1 Nature of damage

Nymphs and adults of this mirid bug suck sap from the leaves, young shoots, inflorescence, developing young nuts and apples. The injury made by the suctorial mouth parts of the insect causes exudation of a resinous gummy substance from the feeding punctures. The tissues around the point of entry of stylets become necrotized due to the action of the phytotoxin present in the saliva of the bug, infesting the tender shoots / inflorescences at the time of feeding. These lesions turn pinkish brown in 24 hours and become black in 2-3 days. Feeding on tender leaves causes crinkling. Affected shoots show long block lesions and may cause die-back in severe cases. Infested inflorescence usually turn black and die, immature nuts may drop off. Heavily infested trees show scorched appearance, leading to the death of shoots and growing tips.

5.1.2.2 Bioecology

The adult bug is reddish-brown, about 6-8 mm long with a black head, red thorax, black and white abdomen. The adults are poor flyers, generally fly below the bush and mating takes place mostly on the undersurface of cashew leaves. Female bug lays eggs in the tender tissues of new shoots, leaf petioles and veins. Presence of chorionic threads projecting outside the tissues is indicative of the presence of eggs inside. They are laid singly and a single female bug lays an average of 50 eggs. The incubation period of the egg is on an average 5-7 days, at a temperature range of 24-32°C and relative humidity of 50-100 per cent. The young nymphs are orange coloured and ant-like. A nymphal period with five instars is of 10-15 days. The female bug lives for about 7 days, while the longevity of male is 9-10 days. The life-cycle is completed in 25-32 days.

5.1.2.3 Seasonal incidence

During rainy season, mosquito bug population is not present in the field but just after rain (December onwards), when the cashew trees are in their most active growth phase with the new flush and tender inflorescence, the condition becomes quite congenial for the multiplication and feeding and during this period this pest rapidly becomes abundant. The pest population reaches its peak, invariable in February-March *i.e.,* in the third month of commencement of the population builds up when the trees are in full blossom. Young trees are liable to suffer more in a longer period than older trees, because of the availability of succulent flush throughout the year. The population build up of the pest is negatively correlated with meteorological factors like minimum temperature, relative humidity and rainfall and positively correlated with sunshine.

5.1.2.4 Management

For efficient management of mosquito bug, it is essential to check the buildup of the pest population on the alternate hosts such as neem and in the vicinity of cashew plantations (Bhat and Raviprasad, 2008). Among the various insecticides tested, profenophos 50 EC 0.05% (@ 1.0 ml/lit) one spray at flushing phase (November-December), chlorpyriphos 20 EC 0.05% (@ 2.5 ml/lit) one spray at flowering phase (December-January) and acephate (@ 1g/lit) one spray at nut formation phase (January-February) were found to be superior in checking the pest.

A through foliar coverage is a must. Avoid indiscriminate use of synthetic pyrethroids as it causes flare-up of sucking pests. If no new growth is seen from the dried up shoots or branches in spite of spraying insecticides in the severely infested cashew trees, such shoots or branches should be split open for confirmation of "die back" disease, which is caused by fungal pathogen (*Botrydiplodia theobromea*).

For controlling fungus infection, pruning of dried shoots or branches, swabbing the cut surfaces with Bordeaux mixture paste (10%) and spraying the tree with Bordeaux mixture (1%) are required in addition to insecticidal sprays (Thirumalaraju *et al.*, 2005). At the out-break situation, the management programme against this pest should be launched on large scale community basis as the efforts made by an individual farmer may not be of much use (Beevi and, Mahapatra, 2007). The eggs of the mosquito bug are naturally parasitized by *Telenoums sp. and Erythmelus helopeltidi* (Sundararaju, 2004). Red weaver ant *Oecophylla smaragdina* is an efficient insect predator which limits both nymphs and adult mosquito bug by patrolling the foliage and active chasing and biting behavior (Peng *et al.*, 1995). Conservation and augmentation of this ant species in cashew plantings is highly beneficial.

5.1.3 Mealybug *Ferrisia virgata* Cockerell (Pseudococcidae: Heteroptera)

The mealy bug, *Ferrisia virgata* is a serious pest of cashew in all cashew growing areas. Two other species of mealy bugs infesting cashew includes *Planococcus lilacinus and Planococcus citri*.

5.1.3.1 Nature of damage

The nymphs and adults colonize on the lower surfaces of tender leaves, twigs, inflorescence panicles and fruit peduncles and suck the sap. The insects being sessile remain stationary while feeding on plant parts. The terminal leaves turn pale yellow and curl downwards due to continuous drainage of sap by large mealy bug populations. Besides causing direct damage, the bugs excrete a sweet and sticky substance. It falls on the upper surface of lower leaves, twigs and fruiting parts. This clogs stomata and makes the plant surface sleek on which the black sooty mould fungus develops. This black coating, covering the upper surface of leaves and twigs impedes the photosynthetic activities. Nut qualities also suffer. Ants are attracted to the sweet excreta and aid in spreading the bugs. Nut yield is lost heavily in affected trees.

5.1.3.2 Biology

The tailed mealy bugs have reproductive potential of laying 100-300 eggs in masses covered with cottony white mealy filaments. The eggs are amber in colour and within a day they hatch into crawlers. Nymphal period lasts for 26-45 days. Adult lives for 15-20 days. The total life period is 45-65 days.

5.1.3.3 Management

Regular monitoring and early detection of the infestation are essential to combat this menace. The plantation and neighbouring areas should be free from weeds and alternate hosts. The infested portion of the plant with mealy bug colonies may be pruned and destroyed. Fallen leaves under the tree canopy should be collected and burned to avoid further spread of the pest. If the pest incidence is widespread over large area, periodic pruning of affected portions coupled with spray of dichlorvas 76 WSC 0.2% (@ 2.5 ml/lit) or profenophos 50 EC (@ 2ml/lit) or dimethoate 30 EC 0.05% (@1.75ml/lit) in combination with fish oil resin soap @ 20 g per litre of water reduce the bug incidence effectively. Several local predators and parasitoids are known to regulate *F.virgata* population in nature. The predatory larvae of chrysopids (green lacewings) *Chrysoperla* sp. are potent predators on eggs and nymphs of the mealybugs.

5.2 Other Pests

5.2.1 Thrips, *Selenothrips rubrocinctus* Giard. (Thripidae: Thysanoptera)

Three species of thrips *viz., S. rubrocinctus, Rhipiphorothrips cruentatus* and *Retithrips syriacus* have been reported as foliage pests of cashew of which *S. rubrocinctus* is of economic importance. This thrips is widespread throughout the tropics; however, it is not considered to be a major pest because its infestation is restricted in small areas.

5.2.1.1 Symptoms of damage

Adult and young thrips are seen in colonies on the lower surface of leaves and suck the sap. As a result of rasping and sucking activities, the leaves of infested young trees become pale-brown and crinkled with roughening of upper surface. In severe cases, there will be shedding of leaves and even stunting of growth. The leaves of post-monsoon flushes reach near full size but not matured, are mostly invaded by thrips but as the leaves mature, the number of newly hatched nymphs declines to accept the old leaf tissue and finally the remaining population disperse. In most of the young plants, thrips are found to appear on the foliage and cause heavy damage, particularly in summer months. The population increases during the dry season from December to January to a peak in April-May and then rapidly declines during wet season.

5.2.1.2 Management

Spraying with imidacloprid 17.8% SL (@ 1.5 ml/lit) at the time of emergence of tender shoots and panicles is very effective for its control.

5.2.2 Leaf and blossom webber, *Lamida moncusalis* Walker (Pyraustidae: Lepidotera)

Leaf and blossom webber, *Lamida* (= *Macalla*) *moncusalis* is reported to be a major pest in East-Coast, particularly in Tamil Nadu, Andhra

Pradesh and Odisha. Another species, *Orthaga exvinacea* Hamps (Noctuidae) has also been recorded as pest on cashew, but it is of minor importance.

5.2.2.1 Symptoms of damage

Leaf and blossom Webbers attack new flushes and inflorescences. The caterpillars of this pest web the shoots and inflorescences together, remain inside and feed on them. Subsequently the webbed portion of the shoots and blooms dry up. Hence, it is called shoot and blossom webbing caterpillar. The galleries of silken webs reinforced with castings and scraps of plant parts are indicative of the presence of caterpillars inside the webbed portion. The incidence is found severe mostly on young trees.

5.2.2.2 Biology

Males of *L. moncusalis* are dark, while the females are olive green. The eggs are laid singly or in small groups of 3-5 on the leaves during night, more usually on the ventral surface and occasionally on the tender shoots, finally glued on to the surface. The incubation period lasts 4 to 7 days. The caterpillar is dark green in colour with yellow longitudinal bands and brownish dorsal lines. Full-grown larva measures 24.0 to 27.0 mm in length and the larval period is 16 to 24 days with five larval instar. Population takes place in cocoons in the leaf webs. The pre-pupal period lasts 1-4 days and the pupal period 9-14 days. Life-cycle is completed in 35 days on cashew.

5.2.2.3 Management

Spraying profenophos 50 EC (@ 2 ml/lit) or chlorpyriphos 20 EC 0.05% (@ 2.5 ml/lit) at flushing period will control this pest effectively.

5.2.3 Hairy caterpillar, *Metanastria hyrtaca* Cramer (Lasiocampidea: Lepidoptera)

Of the several species of defoliating hairy caterpillars recorded as cashew pest, *M. hyrtaca* assumes serious proportions occasionally. Other species include: *Circula trifenestrata* Helfer; *Bombotelia jacosatrix* Guen., *Lymantria* sp., *Thalassodes quadraris* Guen, etc. which cause damage by defoliation on isolated trees in certain localities.

5.2.3.1 Symptoms of damage

The caterpillars are nocturnal feeding generally after sun set. They are gregarious and congregate on the lower surface of leaves. Initially, they feed on fresh leaves starting from petiole and continue half way on a leaf, leaving the midrib intact. In later stages, the caterpillars feed voraciously on mature leaves and tender twigs leading to complete defoliation leaving only the bare branches. After feeding, they hide at the base of the tree trunk, shaded branches and on the ground under dry leaves during daytime. Dark green, pea size fecal pellets littered in mass, on the soil under tree canopy reveals the presence of the pest in the orchard. *M.hyrtaca* is a polyphagous pest known to infest drumstick, jamun, sapota, cocoa and garden plants. The pest appears during early monsoon and continues upto December.

5.2.3.2 Biology

The female moth is fairly bigger (body length 29-31 mm) than adult male (body length 25 mm). Male moths have a black patch with a small white spot in the centre of fore-wing and two angulated transverse lines on either side of the patch, but there are no black patches present in the fore-wings of the females. The gravid moth lays 180-250 eggs in small clusters on the lower surface of cashew foliage or on tender twigs. Eggs are spherical in shape and ash grey to black in colour. The eggs hatch in 7-9 days. The larval period is 30 to 40 days. Pre-pupal period takes 2-3 days and the pupal period ranges from 12-18 days. The total life cycle would be 55-65 days with an average of 60 days from eggs to adult.

5.2.3.3 Management

The caterpillars which hide at the tree trunk and on the ground under dry leaves and debris should be collected and destroyed. This is an easy and effective way of checking the pest. When young larvae are noticed, the trees are to be sprayed with profenophos 50 EC 0.05% (@ 1.0 ml/lit) or chlorphriphos 20 EC (@ 2.5 ml/lit). While spraying, the tree trunk and soil surface should also be treated simultaneously. Thorough coverage of foliage is essential for effective control of the pest.

5.2.4 Apple and Nut borer, *Thylocoptila panrosema* Meyrick (Pyralidae: Lepidoptera)

The apple and nut borer cause 10% of the yield loss, during the years of severe infestation in certain tracts.

5.2.4.1 Symptoms of damage

The young larva of *T.panrosema* move to the joints of nut and apple, scrape the epidermis and then bore into them. In later stages, they bore into tender apples and nuts and feed on them. The borer affected nuts do not developed, they become shriveled and dried up resulting in pre mature fall of nuts and apples. Usually, the borer tunnel near the junction of apples and nuts, and the entry holes are plugged with excreta.

5.2.4.2 Biology

The adult is medium sized moth with dirty black fore-wings and pale dark hind wings, with a wing span of about 15-20 mm in length. There are 5 larval instars which are completed in 15-23 days. The fully grown larvae drop to the ground and pupate in earthen cocoons. The pupal period lasts about 8-10 days.

5.2.4.3 Management

Best control could be obtained by spraying profenophos 50 EC (@ 2 ml/lit) at the time of fruit setting. For effective control of the pest, spraying should always be taken up only during cool hours of the day.

5.3 Tips for effective pest control

- As precautionary measure for mosquito bug, first spray may be given as soon as the new flush is emerged.
- Avoid pruning at active flushing growth period as it induces frequent flushes, which lead to proliferation of many foliar pests. Prune only dried or pest infested dried twigs or branches after harvest.
- Insecticidal spray should be coincided with right phonological stage of tree. While spraying, the nozzle should be directed in such a way as to cover the entire canopy of trees.

- Spray operation should be done in evening hours, which helps in effective utilization and to prevent loss of pesticide through evaporation.
- The time of pesticidal application should be decided after monitoring pest population or its incidence.
- Avoid repeated application of same pesticide as it leads to develop pesticide resistant pest population, and also cause resurgence of sucking pests.

5.4 Disease management

The cashew tree is subject to the attack by a multitude of diseases both in nursery and established plantations. Compared to the number of insect-pests and the intensity of damage caused by them, the magnitude of the disease problem is not that much severe. However, some major diseases and their management are detailed here.

5.4.1 Nursery Diseases

5.4.1.1 Damping-off

Damping off is caused by a variety of seed and soil borne fungi, majority of which are soil inhabitants. These includes, *Pythium spp., Phytophthora spp., Fusarium spp.,* and *Rhizoctonia solani.* Occasionally, *Sclerotinia sclerotiorum, Sclerotium rolfsii, botrytis spp., and Cylindro-cladium* can also cause damping-off. The disease occurs in nurseries where drainage conditions are poor. Affected plants turn pale and show water soaked girdle of darkened tissues around the stem. Such seedlings eventually topple over or rot. In severe cases, the leaves also show water soaked lesions which enlarge and coalesce. Close spacing, continuous nursery growing, cool, wet, highly organic and neutral to basic soils favour damping off diseases.

5.4.1.2 Seedling blight

Seedling blight is a fungal disease. *Fusarium sp., Pythium sp., Phytophthora palmivora, Cylindrocladium scoparium* have been reported as causal agents. The disease occurs in nurseries having ill-drained conditions. The disease is characterized by tiny yellowish sports that appear on leaves of root stock. These pin-head-like spots gradually increase in size. As a result, the surrounding tissues turn brown and then become ash coloured, consequently the leaf tips dry up. Infection also spreads to petioles and leaves. Elliptical lesions appear on petioles and girdle the stem tissues.

5.4.1.3 Management of nursery diseases

- Maintain good plant health and nursery sanitation with proper regulation of watering and drainage throughout the season.
- Precautionary drenching of nursery stocks and grafted plants with Bordeaux mixture (1%) or carbendazim 50 WP @ 1g/lit or copper oxy chloride 50 WP @ 2.0 g/lit or streptocyclin sulphate @ 1g/10 lit of water once in ten days during the rainy months effectively check nursery diseases.
- Collect and destroy the diseased plants by burning to prevent further spread of disease.
- Proper drainage in nurseries is essential.

5.4.2 Anthracnose

Anthracnose is a fungal disease caused by *Colletotrichum gloeosporioides* resulting in necrotic patches on leaves, twigs and fruits. The affected leaves become crinkled, dry up and fall off. The twigs show elongated dark brown to black lesions and ultimately dry up. Dark brown lesions are formed on inflorescence stalk and laterals of the panicle which eventually wither and dry. When immature fruits are infected, these turn black and fall prematurely. High humidity, intermittent rains and temperature regime of 24-32°C during the period of flower formation and fruit setting are most favourable for the spread of this disease. In order to prevent the occurrence of this disease avoid overcrowding of the trees in an orchard and adopt adequate phytosanitary measures.

5.4.2.1 Management

In the initial stage of infection, prune the affected parts and burn the same. This will prevent the spread of the disease. Effective control can be obtained by cutting and removal of the affected plant parts and by spraying Bordeaux mixture 1% or mancozeb 0.2% or copper oxy chloride 0.3%. Infection of the fungus is usually preceded by infestation of tea mosquito bug. So, combined application of insecticide and fungicide are effective and more economical.

5.4.3 Inflorescence Blight

This is a common disease, especially during the monsoon period. The disease is caused by the fungi, *Colletotrichum gloeosporioides* and *Phomopsis anacardii* in association with tea mosquito bug. The disease is characterized by drying up of floral branches. Reddish brown, shiny water soaked lesions develop on the main and secondary rachis. The lesions turn pinkish brown, enlarge and soon turn scabby. Gummy exudation can be seen on the affected regions. The adjoining lesions coalesce and result in the drying up of inflorescence and present a scorched appearance. The nuts and apple show rotting and shrinking. The incidence of the disease aggravated when cloudy weather prevails.

5.4.3.1 Management

A combination spray of a fungicide copper oxy chloride 0.2% and an insecticide monochrotophos 36 WSC 0.05% @ 1.5 ml per litre is recommended.

5.4.4 Gummosis

The disease is caused by *Ceratocystis* sp. The disease is characterized by exudation of a reddish brown liquid from the main stem and branches, which later turn black in colour. A longitudinal crack also appears on the affected region from which gum exudes. The inner tissues of the affected portions will be reddish brown having small cavities filled with reddish fluid. Chisel out the affected part and apply Bordeaux paste or swab the cut portion with suspension of copper oxy chloride.

5.4.5 Pink Disease or Die-back

Pink disease is of minor importance caused by *Pellicularia salmonicolor* (Berk and Brown). It is prevalent during South-West monsoon period. The disease is prevalent during rainy season. Affected branches initially show white patches on the bark and a film of silky thread of mycelium develops on the branches during monsoon season. The fungus penetrates into deeper tissues and interferes with the transport of nutrients. As a result, the shoots and branches wilt and dry up from the tip downward and hence the name die back. Later the fungus develops a pinkish growth which represents the spore mass. In advanced stages, the bark splits and peel off while the leaves turn yellow and fall off.

5.4.6 Management

Management measures for this disease include pruning of the affected branches or shoots below the spot of infection and swab the cut ends with Bordeaux paste. Spraying copper oxy chloride 50WP @ 2.0 g/lit (or) Bordeaux mixture 1% in June-July and again in November-December effectively control the disease.

❑❑❑

Chapter 5

Processing of Cashew Nuts

5.1 Processing

Processing of nuts is the recovery of kernels from raw nuts by manual or mechanical means. India has been a pioneer in cashew processing. The growing demand for kernels in the world market and the availability of cheap labour mainly of women possessing the requisite skills are the favourable factors for the rapid growth of processing industry in the country and it has a monopoly over the manual processing of cashew. Though mechanization is introduced in cashew processing, the availability of skilled and cheap labour in India and better quality of kernels under manual processing, limits the scope for extensive mechanization. Factories in general have mechanization in roasting/boiling and packing. Predominance of manual labour continues in the processes like shelling, peeling and grading. In India processing mostly consist of moisture conditioning, roasting, shelling, kernel drying, peeling, grading and packing. Grading of raw nuts before processing reduces broken kernel. Nuts are conditioned by sprinkling water and allowed to remain moist for 24-48 hours (Nagaraj, 1998).

5.1.1 Cleaning

All raw nuts carry foreign matter, consisting of sand, stones, dried apple etc. The presence of foreign matter in the roasting operation can be avoided by cleaning the nuts. The raw nuts can be sieved by hand using a ¾ inch mesh sieve.

5.1.2 Soaking

The next stage is to soak the nuts in water to avoid scorching them during the roasting operation. This can be done by placing the nuts in a 40-45 gallon drum or vat and filling it with water until all the nuts are covered. After being left to stand for about ten minutes, the water should be drained off via a plug near the base of the drum. The nuts should then be left for periods of not less than four hours in order to allow the water left on the surface of the nuts to be absorbed. The process of covering the nuts with water, draining and standing should be repeated with the same nuts about three times until a moisture content of 9% is reached.

Where the production output runs from 2-10 tons of nuts per day, a simple cleaning and conditioning arrangement can be used. Two people open the sacks of harvested nuts on a stand and clean the raw nuts as they are moved along a flat sieve, to two vats which are used for storage until the soaking process begins. Two vats are useful because one can be emptied while the other is being filled.

5.1.3 Roasting

Roasting makes the shells brittle, besides making the extraction of kernels easier. A slight under roasting or over roasting adversely affects the quality of and recovery of kernels. In the open pan roasting method, one kilogram of nuts are kept in shallow iron pans or earthen pots and are heated over an open fire. The nuts are rapidly turned to prevent charring. During the process of roasting, large quantities of shell liquid and smoke would come off and cause an irritation and injurious effect on the skin of personnel engaged in the operation. The roasted nuts are then removed from the pan and thrown on the floor. They are quickly covered with earth which would absorb shell oil adhering to the roasting nuts and also cool them. The nuts are then subjected to subsequent operations. The cashew nuts are also roasted by drying under sun for two to three days, when they lose much of the moisture contents and become brittle enough for shelling. The other improved methods of roasting cashew nuts are drum roasting, oil bath roasting and steam boiling (Balasubramanian,1998).

5.1.3.1 Drum roasting

In drum roasting, the nuts are fed into an inclined rotating drum that is heated initially to such an extent that exuding the oil ignites and burns, thus charring shell. Drum maintains its temperature due to burning of CNSL oozing out of nuts. Roasting takes 3-5 minutes. Roasted nuts which are still burning are covered with ash to absorb oil on surface. Shell becomes brittle and out turn of whole kernels is reported the highest in the method.

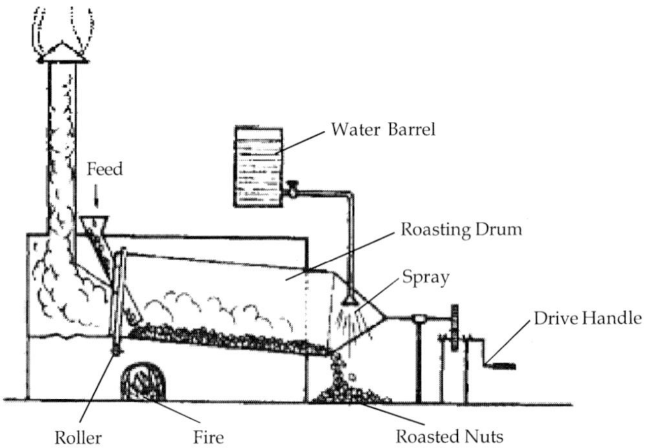

Fig. 12. Drum roaster for cashew nut roasting

5.1.3.2 Oil bath roasting

Oil bath roasting is a traditional method followed in a few processing factories of Kerala and Karnataka. In this method, raw nuts are passed for 1-3 minutes through a bath of CNSL maintained at a temperature of 190-200^0 C by means of a screw or belt conveyor. Roasting equipment consists of a rectangular vessel 60-90 cm deep with a flat bottom. Whole assembly is embedded in brickwork furnace which uses spent cashew shell as fuel.

5.1.3.3 Steam roasting

The steam roasting process, commonly known as "cooking process" consists of a baby boiler followed by a steam cooker where the cashew nuts are cooked with steam. Steam boiling is adopted in factories where hand-and-leg-operated shelling machines are used. Nuts after conditioning are given a mild roasting in a closed vessel for 20-25 minutes at 100-120 PSI to loosen kernels from shell and make its removal easy.

5.1.3.4 Baby Boiler

The baby boiler is a hand stoked fire tube boiler, which produces steam at 7.0–8.5 kg/cm^2. The roasted cashew shell or de-oiled cake are fed manually (about 2 to 3 kg fuel once in 15 to 20 minutes) at the bottom of the boiler. The combustion air is drawn through the grate by natural draft stack at the top of the boiler. The flue gases from the boiler are discharged into atmosphere through the stack 12–15m high.

5.1.3.5 Cashew Nut Cooker

A cylindrical steam cooker with provision of cashew nut feeding at the top and discharging of cooked nuts from the side near bottom, has a capacity of holding 4 bags (80 kg each) of cashew nut in a batch. Once the cashew nut is loaded, steam from a boiler is introduced into cooker at a pressure of 7.0–8.5 kg/cm^2. The cashew nuts in the cooker are steam cooked and when all the nuts are sufficiently cooked, the excess steam starts releasing near the bottom outlet. The steam is injected into the cooker till the steam starts escaping from the outlet mouth of the cooker. This process takes about 10–20 minutes time. Then the steam injection into the cooker is stopped and the condensed water at the bottom of the cooker is discharged in to a container and disposed on ground through septic tank . In cashew nut cooking process in Kerala have different process in steam injection quantity. The steam is injected for about 15 minutes even after the steam starts escaping from the outlet mouth of the cooker. This method consumes more steam.

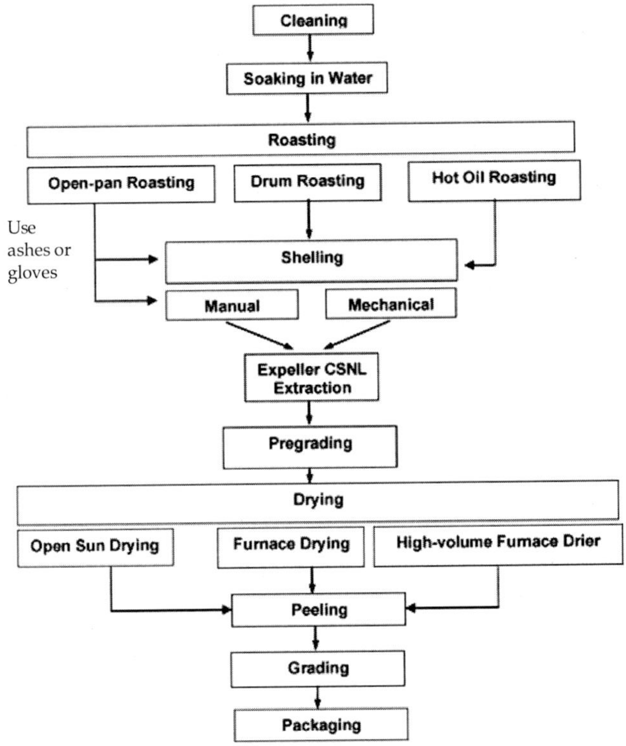

Fig. 13. Process flow chart for cashew nut processing

The cooked cashew nuts are removed from the bottom of the cooker and spread on the floor for cooling. The cooled nuts are sent for cutting section to cut open and collect the kernel. The total batch time is about 40–45 minutes. Usually, the units are provided with a baby boiler and minimum two cookers, while one cooker is in cooking operation, the other cooker is in unloading and loading operation. Normally the cooking operation is carried out for 2–4 hrs in a day, cooking 5 to 10 batches of nuts, depending on the production capacity of the units.

5.1.4 Shelling

The objective of shelling is to produce clean, whole kernels, which are free of cracks. Shelling has always been manually performed in India. Other countries have difficulty in competing with the great skill of Indian workers. India has therefore enjoyed a virtual monopoly of cashew processing for a long time. Shelling is a manual process that entails putting the nut against one sharp blade and bringing another blade, which is on a foot powered lever, through the outer shell. The blade on the foot lever is raised by an enthusiastic stomp allowing the outer shell to separate from the nut. The nut inside is carefully picked out of the shell using a nutpick. After shelling, shell pieces and kernels are separated, and the unshelled nuts are returned to shelling operation. Output is 15-20 kg kernels in 8 hours with 90% whole kernels. Mechanical shelling can also be done where shell is cut and kernel is scooped out using a sharp needle. Output of a worker is 14-22 kg of kernels. Main disadvantage is that kernels are contaminated by CNSL from mildly roasted nuts. Varying size of nuts requires careful manipulation during cutting to avoid injury to hands.

5.1.5 Separation

After shelling, shell pieces and kernels are separated and the unshelled nuts are returned to the shelling operation. Usually blowers and shakers are used to separate the lighter shell pieces from the kernels. The greatest problem is to recover small pieces of kernel sticking to the shell. This is usually done manually from a conveyor belt used to carry all the sorted semi-shelled nuts.

5.1.6 Pre-grading

Pre-grading can be done before or after drying the kernels and may greatly reduce the final grading work. For large-scale processors pre-grading can be done mechanically, separating mainly the whole from the broken kernels and sometimes separating the different size groups of whole kernels.

5.1.7 Drying

The shelled kernel is covered with the testa, the removal of which is facilitated by drying the shelled kernel, to produce the blackened kernel. Drying helps to reduce the moisture content 2-4 % in kernels. Drying causes shrinkage of the kernel, thereby allowing the testa to be easily removed either mechanically, or by hand with a knife. Drying also protects the kernel from pest and fungal attack at this vulnerable stage. Drying usually takes six - eight hours, at a temperature of around 70^{0}C-75^{0}C. A uniform temperature throughout the drier is essential to avoid under-drying or scorching. A tray dryer is used, for drying cashew kernels. The dryer contains a series of mesh- bottom trays that are slotted into the drying cabinet. The trays are of a size that can be lifted when full. Hot air circulates over the trays and is exhausted through the chimney. Cashew oil cakes are burned to provide the heat source.

5.1.7.1 Kernel Drying (Borma)

The kernel coming from the shelling section (Roasting Process) as well as Cutting Section (Cooking Process) contains a brown cover, known as 'testa', over it. To remove testa over the kernel and also control the moisture content in the kernels, they are exposed to prolonged and controlled heating with hot air at 80-90°C for about 6-7 hrs in the chamber. About 5–8% of moisture is removed from the kernels in the process. This process is known as "Borma". Indirect heating of atmospheric air generates the hot air over hot surface in a "Oven". The Borma Ovens are fired with cashew shell (from Roasting Process) or de-oiled cashew cake (from Cooking Process) for about 4–6 hrs depending on the quantity of kernels to be dried (Fig.1 4). The flue gases about 13–15m high. In Kerala and Maharashtra most of the small scale units carried out the Borma operation by electric heaters which consume large electricity and the process will be dependent on power supply.

5.1.8 Peeling

At this stage, the testa is loosely attached to the kernel, although a few kernels may have already lost the testa during prior operations.

Those parts still attached to the kernel, can be removed by scraping of the testa with a blunt knife, which is the most effective way of removing it. It is essential that the testa is removed with minimum pieces. It is also essential that the entire testa is removed. The testa is gently scrapped with a blunt knife and removed. Average peeling capacity is 7-10 kg/head.

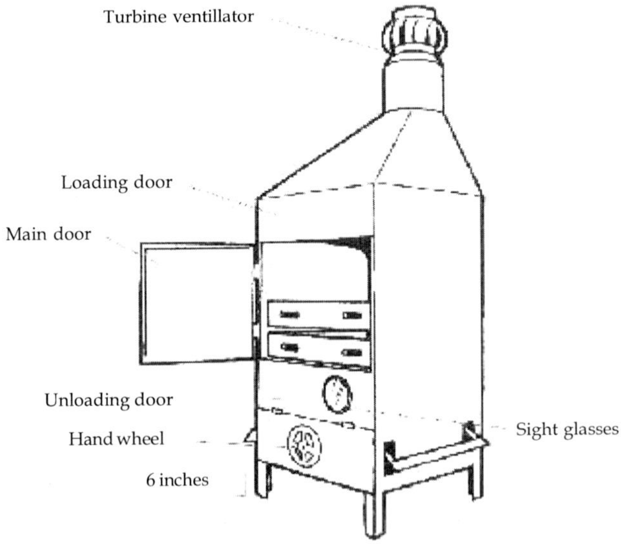

Fig. 14. Borma Oven for cashew kernel drying

At the end of the day, the removed testa is winnowed and all cashew pieces removed. The dust and very fine pieces that cannot be peeled, together with the diseased pieces, are classified as refuse. The brown kernels, which are not been separated out during the shelling operation, are removed and discarded. Strict cleanliness in the peeling operation is observed in the peeling room and its facilities, by all personnel. All workers follow basic codes of hygiene. The peeled kernels are vulnerable to insect infestation and mould growth. They are also prone to rodent attack and are stored in rodent-proof containers or rooms.

5.1.9 Grading

In order to safeguard and guarantee quality, producers and exporters have introduced quality standards which must be met by cashew exporters. Kernels are graded based on size and colour according to specifications prescribed by Government of India under

Export Acts 1963. Cashew kernels are selected on the basis of the number per unit weight, in accordance with the weight of the kernels. They are also classified either as wholes, chips, splits, butts or baby bits, in accordance with the integrity of the kernel. White or ivory kernels are preferred over brown ones. There is a maximum permitted moisture level (both for raw cashews and cashew kernels) and the product must be free from insects, mould, rancidity and extraneous materials. The highest price is paid for better quality kernels of the W180 and W210 grades which are the largest and heaviest grades.

The grading operation is important since it is the last opportunity for quality control of the kernels. After the kernels are extracted from the shells, dried and peeled, they are graded for export according to size and condition. The grading system is known as the American Standard, which is also incorporated in the Indian Government export criteria. Kernels are categorized on the basis of colour and condition. Peeled cashew nuts can be classified into between 11 and 24 grades. These are roughly divided into three groups: white wholes, white pieces and scorched grades.

5.1.9.1 White wholes

- W180 (super Large) Between 120 and 180 kernels per lb (266-395 per kg)
- W210 (large) Between 200 and 210 kernels per lb (395-465 per kg)
- W240 Between 230 and 240 kernels per lb (485-530 per kg)
- W280 Between 270 and 280 kernels per lb (575-620 per kg)
- W320 Between 300 and 320 kernels per lb (660-706 per kg)
- W450 Between 400 and 450 kernels per lb (880-990 per kg)

5.1.9.2 White pieces

- Butts a kernel broken cleanly across the section of the nut.
- Splits a kernel which has broken down the natural line of cleavage to forma cotyledon.
- Pieces a kernel which has broken across the section but does not qualify for a butt and is above a specific size.
- Small pieces as above but smaller.
- Baby bits very small pieces of kernel which are white in colour.

5.1.9.3 Scorched grades

Wholes Whole kernels that have been slightly scorched during the process but are otherwise sound. These are not graded according to size.

- Butts that have been scorched.
- Splits that have been scorched.
- Pieces As for pieces, but which have been scorched during processing and contain all but the very small pieces.

Throughout the grading of white whole kernels, the weight must be constantly checked. This is done with the use of a small counter scale with 250g (½ lb) of graded kernels accurately weighed out and counted. From this, the count per kg (or per pound) can be calculated. Quality has emerged, ahead of price, as the most vital criterion for any item if it is seeking entry into the global market. Quality aspects include safety, reliability, durability and acceptability of the product to the consumer. Small-scale processors have to match the standards set by importers, consumers and standards agencies. Quality assurance procedures are an essential element of any processing operation to ensure product consistency between different batches.

5.1.10 Rehumidification

Before the kernels are packed it is necessary to ensure that their moisture content rises from 3% upto around 5%. This is to make the kernels less fragile, thus lessening the risk of breakage during transport. In humid climates, the kernels may absorb enough moisture during peeling and grading to make a further rehumidification process unnecessary.

5.2 Packing

The normal packaging for export of kernels is in air-tight tins of 25lbs in weight. The packing needs to be impermeable as cashew kernels are subject to rancidity and go stale very quickly. The tin will be familiar to most tropical countries as it is a replica of the four gallon kerosene or paraffin oil tin. If possible the tins are made locally as movement of empty tins overseas is expensive. Alternatively, it might be arranged to purchase components and finish the manufacturing locally. This may be done by arrangement with tin manufacturers. The output of a tin manufacturing line is usually too large for one consumer but some cashew nut processors have in fact installed their own tin making plant and supply other processors.

After filling and weighing, the cap should be soldered on in preparation for the 'vita pack' process. This consists of removing all air from the tin and substituting this with carbon dioxide (CO_2). The advantages of packing cashew kernels in carbon dioxide are two fold. Firstly, carbon dioxide is an inert gas and will not support life. Any infestation that may have been present is therefore arrested. Secondly, carbon dioxide is soluble in cashew oil and goes into solution as soon as the seals are made. In a short space of time, it can be seen that a decrease in pressure takes place as the carbon dioxide goes into solution and the sides, top and bottom are drawn inwards. Thus, the kernels are held tight in the tin, preventing movement and breakage during transport. Carbon dioxide, being a heavy gas causes the upward displacement of air and will remain in the tins after the filling process. Some large-scale machines will operate on six tins at a time, creating a vacuum in each and then filling with carbon dioxide.

Some processors do not have vacuum pumps and displace the air in the tin by feeding in carbon dioxide through a small hole in the bottom of a side of the tin. The carbon dioxide valve is turned off when all the air has been replaced. The holes in the tin are then sealed, with the hole at the bottom of the side of the tin being done first, and the one on the top last.

Cashew kernels are offered to US and EU in food grade flexi packs wherein the kernels are packed under vacuum with CO_2 and N_2 in the ratio of 60:40. This is the most accepted packing of cashew kernels, as the buyers don't have to bother much to dispose off the cans after taking out the kernels.

5.3 Infestation

Far too little attention is paid to the infestation hazards of cashew kernels. These hazards are more prevalent at some times of the year than others, however, a good processor will be vigilant all the time.

The main insect pests are:

- Ants
- Grain wevils
- Meal moths

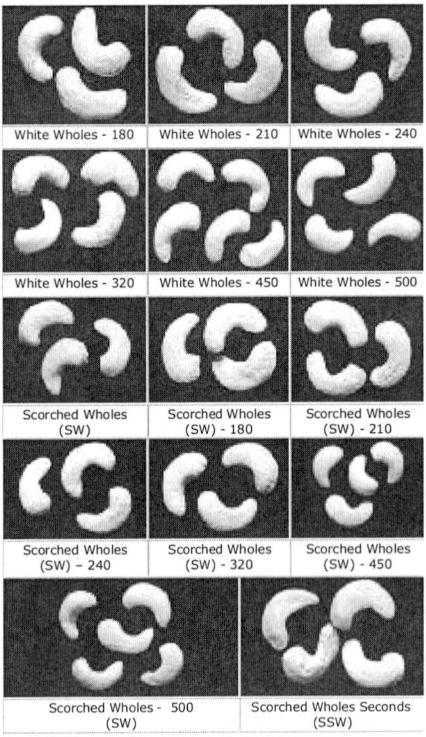

Fig. 15 a. Different grades of cashew

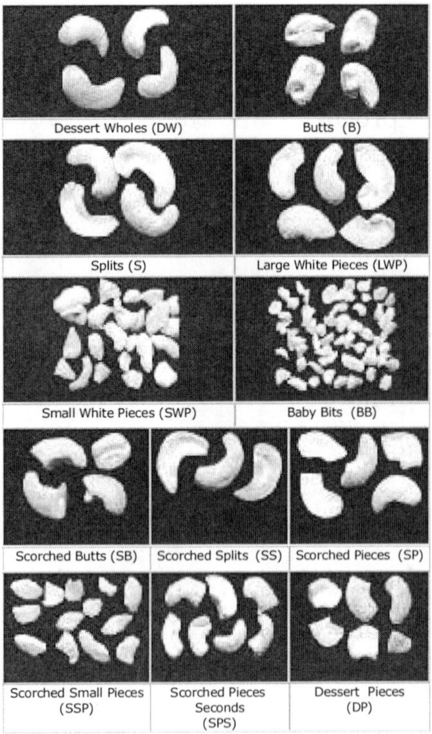

Fig. 15 b. Different grades of cashew

The most important defence against infestation of any type is cleanliness and is essential in the rooms used for drying, peeling, grading, conditioning and packaging. Floors and walls must be sound and free from cracks. They should be kept white-washed regularly. Some processors have filled the corners and places where the wall meets the floor with a curved filling so that the room can be properly swept, all corners having been eliminated. Speed of operations between drying and packaging must be stressed as this reduces the critical period when attacks may occur to a minimum. The equipment used must also be thoroughly cleaned on a regular basis as insects may breed in hidden crevices and gaps.

5.4 Procurement of Raw Nuts

Procurement of raw nuts is an important operation for the cashew processing units. The sample cashew processing units in Kerala procured locally available raw nuts and imported nuts from countries like Tanzania. 60 per cent of the sample cashew processing units in

Orissa was importing raw nuts from African countries like Guatemala, Zambia, Ivory Coast, etc. Cashew nuts were also procured from neighbouring states of Andhra Pradesh and Chhattisgarh apart from procuring the same from the local areas. The average price of raw cashew nuts for the sample units was Rs. 35 per kg. The sample processors without boilers including the sample SHGs in Tamil Nadu were procuring raw nuts only from the domestic market (Perambalur, Puducherry, Nellore and Panruti) whereas; the sample processors with boilers were procuring imported raw nuts through Tuticorin port besides procuring raw nuts from the above mentioned places. The average cost of procuring raw nuts in the domestic market and imported nuts by the sample processing units in Tamil Nadu is depicted in (Table 11).

Table 11. Average Cost of Domestic / Imported Raw Nuts in Tamil Nadu (Rs. Per bag*)

Particulars	Domestic Raw nuts	Imported Raw nuts
Purchase price of raw nuts	2700	2400
Brokerage	25	24@
Purchase Tax @10%	270	-
Transport cost	12	81.81**
Labour cost for loading/unloading	5	5
Total	3012	2510.81

*1 bag = 80 kg @ 1% of Realizable Value ** From Tuticorin Port
Source: NABARD, 2010.

The average cost per bag of domestic and imported raw nuts were Rs. 3012 and Rs. 2510.81 respectively. The difference in the cost was mainly due to the difference in the purchase price of the raw nuts. The size and quality of the local variety was considered to be relatively better by the processing units as they were larger and whiter. Ninety percent of the purchase cost was towards payments for domestic raw nuts and the same was 95 percent for imported nuts. Remaining cost incurred was towards brokerage, purchase tax, transport cost and loading/unloading.

5.5 Economics of Cashew Processing

The economics of cashew processing per bag of raw nuts (80 kg) is depicted in Table 12. The sample processors obtained a total of 23 kg of processed kernels, 55 kg of shells and 2 kg husk from a bag of raw nuts of 80 kg. The shells were sold to the CNSL units @ Rs. 2 per kg (Odisha) and Rs. 2.72 per kg (Tamil Nadu). Similarly, the husk was sold to the traders @ Rs. 8 per kg in Tamil Nadu and the same was used to mix in tea leaves. The sample units in Odisha and Kerala were not making regular sale of the husk. The Financial Rate of Return (FRR) worked out to 35 per cent for the sample units in Kerala and the same was (> 50%) in Odisha and Tamil Nadu. This variation was due to the difference in the costs and processing capacity, which was an average of 2400 MT in Kerala, 818 MT (Odisha) and 160 MT (Tamil Nadu) (Table 12).

Table 12. Economics of Cashew Processing (Rs. per bag of 80 kg)

Particulars	Kerala	Odisha	Tamil Nadu
Average processing cost*	735.5	977	374.5
Cost of raw nuts	3455.5	2800	2761.4
Total cost	4191	3777	3135.9
Gross income (including cashew kernels, shells, husk)	4351	4166	3541.25
Net income	160	389	405.25
Financial rate of return (%)	35	>50	>50

*Include costs on boiling/roasting, electricity, labour charges for shelling, peeling, grading, etc. *Source:* NABARD, 2010.

50 per cent of the sample processing units in Tamil Nadu had boilers and drier and 50 per cent had only drier and were getting the raw nuts boiled at other units having boilers @ Rs. 25 per bag. The sample processors without boilers were procuring only domestic raw nuts. Average processing cost per bag of raw nuts comprising the hired boiling was Rs. 348.18; net income was Rs. 181 per bag and the FRR worked out to 35 per cent. This variation with the processors having boilers was due to the difference in the number of bags of raw nuts processed per month, which were 200 and 100 bags per month for the sample processors with and without boilers respectively.

5.6 Employment Generation in Cashew Processing

The processing of cashew in India is a highly labour intensive activity. Employment generation in cashew processing units was 11 person days in Odisha and 7 person days in Tamil Nadu and Kerala for processing 1 bag of raw nuts of 80 kg and was mainly (90%) for women. This was because women were considered to be hard workers and more reliable. Moreover, the work in the processing units involved lesser physical labour, in comparison to agricultural labour. Since the share of women in this sector is large, the growth of this sector will contribute towards women's employment (Balasubramanian, 1998).

5.7 Risks in Cashew Processing and Mitigation Strategies Adopted

Cashew processing involved a number of risks at various stages of operations and the processing units including those owned by SHGs took necessary steps to mitigate most of them. Primary risk was related to the consistent procurement of raw nuts and stock piling of sufficient raw nuts to operate the units on a regular basis. The raw cashew nuts were purchased at the time of harvest from the local market, procured from nearby districts/states and were also imported.

Two major considerations in the decortications of cashew nuts involve avoidance of contamination of kernels by the toxic CNSL and minimization of kernel's breakage. Second risk pertained to difficulties in shelling cashew nuts, which was due to the irregular shape of the nut, the tough leathery outer shell, and the CNSL within the shell that must not be allowed to contaminate the kernel or burn the hands of the worker/s during its removal from the shell. The processing units including the SHGs fine-tuned the process by employing skilled workers in order to achieve quality kernels.

Cashew kernels were expected to have moisture content of not more than five per cent. This risk was avoided by employing a permanent labour to operate the electric borma efficiently and storing in clean and dry place with sufficient protection and packed with utmost care in tins to preserve their quality.

Quality including safety, reliability and acceptability of the product to the consumer has emerged, ahead of price, as the most vital criterion. The sample processors made efforts to match the standards set by the local export houses.

The prices of cashew nuts were subjected to market fluctuations that affect the small processors. The sample SHG units in Tamil Nadu managed this price risk and procured sufficient raw nuts by using a part of the loan amount and their savings as working capital.

The SHG units were situated in the DRDA and Panchayat Building and had a godown, which was used for storing raw nuts as well as graded kernels. The SHGs faced a risk of running their units if the facility of godown and premises was withdrawn by the concerned Government department and the Panchayat.

5.8 Value added products from kernels

Value addition in cashew can be done through preparing sweetened and flavoured cashew from cashew kernel baby bits. Cashew kernel baby bits could be coated with combination of different colours (apple green, chocolate brown, kesari, lemon yellow, orange red and raspberry red) and flavours (vanilla, cardamom, ginger and clove) (*Source:* Cashew Vision 2025, NRCC, Puttur).

Lower - grade kernels are processed into cashew flour, which has a high protein content and is easily digested. Cashew kernel oil is also called as Caribbean oil. Lower grade kernels, which contain 35 to 40 % oil are processed into kernel oil, which is high quality edible oil that is a comparable to olive oil.

Cashew kernel butter is obtained from kernel residue after extraction of kernel oil is used to produce, which is similar to peanut butter. Cake remaining after extraction of oil serves as animal feed. Central Food Technological Research Institute (CFTRI) at Mysore perfected technology for extraction of cashew butter.

The value added cashew products in Cuddalore included salted and roasted kernels with chilly or pepper flavours. The average cost incurred per kg for the kernels was Rs. 220 and was being sold at Rs. 300.

The export of value added cashew kernels from India is insignificant. This is mainly due to the reason that the importers and packers in the major markets like United States do not want the Indian suppliers to send value added products, which they consider, would adversely affect their packing industry. However, there is a scope for increasing the export of value added cashew kernels in the non-traditional markets like West Asian countries.

5.9 Nutritive value of cashew kernels

Table 13. Nutritive value of cashew kernels

Principle	Nutrient Value per 100 g	Percentage of RDA
Energy	553 kcal	28%
Carbohydrates	30.19 g	23%
Protein	18.22 g	32.5%
Total Fat	43.85 g	146%
Cholesterol	0 mg	0%
Dietary Fiber	3.3 g	8.5%
Vitamins		
Folates	25 µg	6%
Niacin	1.062 mg	6.5%
Pantothenic acid	0.864 mg	17%
Pyridoxine	0.417 mg	32%
Riboflavin	0.058 mg	4.5%
Thiamin	0.423 mg	35%
Vitamin A	0 IU	0%
Vitamin C	0.5 mg	1%
Vitamin E	5.31 mg	35%
Vitamin K	4.1 µg	3%
Electrolytes		
Sodium	12 mg	1%
Potassium	660 mg	14%
Minerals		
Calcium	37 mg	4%
Copper	2.195 mg	244%
Iron	6.68 mg	83.5%
Magnesium	292 mg	73%
Manganese	1.655 mg	72%
Phosphorus	593 mg	85%
Selenium	19.9 µg	36%
Zinc	5.78 mg	52.5%
Phyto-nutrients		
Carotene-β	0 µg	–
Crypto-xanthin-β	0 µg	–
Lutein-zeaxanthin	22 µg	–

(*Source:* USDA National Nutrient data base)
http://www.nutrition-and-you.com/cashew_nut.html

5.10 Health benefits of Cashew nuts

- Cashews are high in calories. 100 g of nuts provide 553 calories. They are packed with soluble dietary fiber, vitamins, minerals and numerous health-promoting phyto-chemicals that help protect from diseases and cancers.
- They are rich in "heart-friendly" monounsaturated-fatty acids like oleic, and palmitoleic acids. These essential fatty acids help lower harmful LDL-cholesterol while increasing good HDL cholesterol in the blood. Research studies suggest that Mediterranean diet which is rich in monounsaturated fatty acids help to prevent coronary artery disease and strokes by favouring healthy blood lipid profile.
- Cashew nuts are abundant sources of essential **minerals.** Minerals, especially manganese, potassium, copper, iron, magnesium, zinc, and selenium are concentrated in these nuts. A handful of cashew nuts a day in the diet would provide enough of these minerals and may help prevent deficiency diseases. **Selenium** is an important micronutrient, which functions as a co-factor for antioxidant enzymes such as Glutathione peroxidases, one of the most powerful antioxidants in the body. **Copper** is a cofactor for many vital enzymes, including cytochrome c-oxidase and superoxide dismutase (other minerals function as co-factors for this enzyme are manganese and zinc). **Zinc** is a co-factor for many enzymes that regulate growth and development, gonadal function, digestion, and DNA (nucleic acid) synthesis.
- Cashews are also good in many essential vitamins such as pantothenic acid (vitamin B5), pyridoxine (vitamin B-6), riboflavin, and thiamin (vitamin B-1). 100 g nuts provide 0.147 mg or 32% of daily-recommended levels of pyridoxine. Pyridoxine reduces the risk of homocystinuria, and sideroblastic anemia. Niacin helps prevent "pellagra" or dermatitis. Additionally, these vitamins are essential for metabolism of protein, fat, and carbohydrates at the cellular level.
- Further, the nuts are also hold a small amount of **zea-xanthin,** an important pigment flavonoid antioxidant, which selectively absorbed into the retinal macula lutea in the eyes. It is thought to provide antioxidant and protective UV ray filtering functions and helps prevent age-related macular degeneration (ARMD) in the elderly.

❑❑❑

Chapter 6

Value Addition of Cashew Apple

6.1 Cashew Apple Processing

Cashew apple, the pseudo fruit or false fruit is a by-product of the cashew industry. Cashew apple juices contains sugar, protein, acid and large amounts of vitamin C. The juice also contains an astringent principle, polyphenolic in nature. The juice, though one of the richest source of vitamin C, is not acceptable as a soft drink on account of its high astringency. A number of processes have now been developed for converting the cashew apple into various products such as juice, jam, syrup, chutney and beverage.The ability of cashew apples to supply and fortify the nutritional requirement for vitamin C, particularly in Africa was reported by Akinwale (2000). The author carried out a physico-chemical analysis of some tropical fruits and compared them with those of cashew apple. Cashew apple juice was found to contain the highest amount of vitamin C (203.5mg/100ml.) of edible portion and when cashew apple was blended with other tropical fruits it boosted their nutritional quality. Cashew apple is normally not eaten due to its high astringency. The astringency is due to the presence of tannins, which can be removed by addition of suitable additives before processing.

In certain parts of India like Goa, cashew apple is used to distil cashew liquor referred to as feni. But in the main cashew producing areas of India, 95 per cent of cashew apple is not consumed, as the taste is not popular. Although, KVK, Vridhachalam in Tamil Nadu had been organizing various training programmes for the processing of cashew apple, the sample farmers were not making use of cashew apple in any form of processing. Instead, it was mostly used as a manure or fodder. Similarly, in Koraput district (Odisha), little efforts had been made for value addition to cashew apple and the apples were discarded at the time of collecting raw nuts. Even in Kerala, cashew apple processing had not been taken up on a large scale, except for Madakkathara research station of KAU and very few processors in Kannur.

Juice can be extracted from ripe cashew apples after washing using screw press/basket press/by hand press and about 50-60% raw juice with 9-10% soluble solids could be obtained. An improved hydraulic type juice expeller is capable of crushing an average of 150 kg apple per hour with 80% extraction efficiency. The components responsible for astringency can be removed by adopting the following methods:

- Steaming of cashew apple followed by thorough washing of the steamed fruit in water. Pressure of steam and time of exposure vary from 2 to6 kg and 5 to 15 minutes respectively, according to the quality of fruit and the product to be prepared.
- Immersing the fruit for 4-5 minutes in boiling solution of common salt (2-3%) followed by washing in water.
- Mix gelatin @ 0.5 g/kg of raw juice and allow settling. Decant the clear juice and discard the sediment (gelatin may be dissolved in water by heating).
- Mix polyvinyl pyrolidone @ 1.4 g/kg of raw juice. Allow to settle and decant the clear juice.
- Mix about 125 ml of fresh rice boiled water (gruel) and allow to settle. Decant the clear upper layer and repeat the process using 125 ml of rice gruel.

In order to prevent spoilage of the raw juice, potassium metabisulphite (KMS)@ 2g/kg and citric acid @ 5 g/kg may be added along with the clarifying agent. The clarified juice can be stored for further use. Standard products like syrup, squash, ready-to-serve beverages, etc. can be made using the clarified juice as per specifications. (Pushpalatha and Kutty, 2005). Technologies have been developed at Kerala Agricultural University for utilization of cashew apple for manufacture of different types of fruit products (Josemathew and Mini, 2008).

6.1.1 Cashew apple syrup

The items required for preparation of cashew apple syrup

Fully ripe cashew apple	250kg
Poly Vinyl Pyrolidone (PVP)	10g
Sodium benzoate	6g
Sugar	10kg
Citric acid	150g

The juice is extracted, strained through muslin cloth and clarified by adding 1.4 g of PVP per litre of juice. The mixture is stirred for two minutes after which it is strained again through muslin cloth. Sugar is added at the rate of 1.00 to 1.25 kg for every litre of juice. Sodium benzoate is to be dissolved in a small quantity of warm water before adding to the mixture. All the ingredients are mixed thoroughly and kept as such for three to five hours so that clear syrup forms a separate layer, which can be easily siphoned. After bottling, keep it in a cool dry place. Dilute the syrup five times its volume with plain water to use it as fresh drink.

6.1.2 Cashew apple squash

The items required for preparation of cashew apple squash

Cashew apple	25kg
Poly Vinyl Pyrolidone (PVP)	10g
Sodium benzoate	6g
Sugar	3kg
Citric acid	100g

As in cashew apples syrup preparation, the juice extracted, strained through muslin cloth and clarified by using PVP. After straining, 0.4 kg sugar, 12.5 g citric acid and 0.75 g of sodium benzoate are added per litre of the clear juice. All the ingredients are mixed thoroughly and bottling can be done as described for juice.

6.1.3 Cashew apple juice

The items required for preparation of cashew apple juice

Cashew apple	25kg
Poly Vinyl Pyrolidone (PVP)	10g
Sodium benzoate	6g
Sugar	500g
Citric acid	8g

The juice, after extraction, is strained through muslin cloth and clarified by adding 1.4 g of Poly Vinyl Pyrolidone (PVP) per litre of juice. Stir the mixture for two minutes after which it is strained again through muslin cloth. Boil the juice after adding 60 g sugar. The finished product may be chemically preserved by using sodium benzoate at the rate of 0.08 g and citric acid 0.01 g per 100 ml to provide acidity. The clear juice could be poured into well-sterilized bottles, cork air tight and stored in cool dry place.

6.2 Pulp products

Preparation of value added products from fresh cashew apple like jam, pickles, chutney and candy after removal of astringent principles are given below.

6.2.1 Cashew apple jam

The items required for preparation of cashew apple jam:

Cashew apple (Pulp)	1 kg
Sodium benzoate	0.01 g
Sugar	0.75 g
Citric acid	0.05 g

Fresh and fully ripened cashew apples are used for preparing jam. Cashew apples are washed thoroughly to remove any adhering dust and dirt and immersed in 3% salt solution for three days to reduce the tannin content, after which the fruits are steamed for 15 to 20 minutes at 0.7 to 1.05 g kg steamed pressure. Then the apples are made into pulp and mixed with 750 g sugar per kg of pulp and cook the mixture. A pinch of citric acid is added towards the end of the cooking process to improve the taste. The finished hot jam is stored in well sterilized glass bottles covering with a disc of waxed, tissue paper or molten paraffin wax on the surface of the jam after cooling.

6.2.2 Cashew apple chutney

The items required for preparation of cashew apple chutney

Cashew apple slices	3 kg
sugar	750 g
Salt	178 g
Onion(peeled and chopped)	84 g
Cumin powder, Black pepper, Cardamom powder	21 g
Ginger green	42 g
Cinnamon (coarsely powdered)	7 g
coriander	7 g
Acetic acid glacial	20 ml

Wash the cashew apple after three days of salt treatment just as jam preparation. Remove the undesirable portion and slice them. Chutney is prepared using 750 g sugar, one large sized onion, 42 g ginger, one teaspoon each of cumin seed, pepper, cardamom, cinnamon and coriander powder, salt to taste and 20 ml vinegar for everyone kg of fruit slice. Tie all the powdered spices in clean thin piece of cloth, make syrup of the sugar by adding equal quantity of water. Add the sliced apples, chopped onion, grated ginger, vinegar and salt to it. Drop the spice bag in syrup just a little before the final stage of boiling. Boil the mixture until it is sufficiently thickened and store in sterilized jars.

6.2.3 Cashew apple pickle

The process of preservation of green fruit in common salt or vinegar is called pickling. Raw green cashew apple can be made in to pickle.

The items required for preparation of cashew apple pickle

Cashew apple slices	65 kg
Salt	12 g
Chilli powder	5 g
Coriander	5 g
Fenugreek	2 g
Mustard powder	5 g
Turmeric	1 g
Citric acid	1.5 g
Acetic acid glacial	1 ml
Jaggery	5 g
Gingelly oil	5 ml

Raw green fruit is washed, sliced and kept in 5% salt solution for three days by changing the solution daily. On the fourth day, salt water is removed and pickle is then prepared in the usual way, using 50g chilli powder, 100 ml gingelly oil, 20g fenugreek powder, 100 asafoetida, 5g turmeric powder, 10 g garlic, mustard powder, a pinch of sodium benzoate and salt to taste for everyone kilogram of sliced apple. Asafoetida, dissolved in hot water, added to the boiling gingelly oil. Powdered turmeric, fenugreek, chilli and mustard are added to it. Cashew apple slices are added after a thorough stirring. Citric acid, salt and sodium benzoate are dissolved in hot water. Mix all ingredients well. Transfer the pickle into clean dry glass jar and store the product for one week before use.

6.3 Cashew Nut Shell Liquid (CNSL)

Cashew nut shell liquid (CNSL) is a naturally occurring phenol which is contained in soft honey comb mesocarp of shell. CNSL is a viscous, oily or balsam like substance, pale yellow to dark brown, having bitter taste and caustic properties. Fresh CNSL contains 90% anacardic acid, which is converted into cardanol on heating. Rest 10% is cardol which is mainly responsible for vesicant property. About seven kg crude oil can be extracted from 100 kg shell. Cashew Nut Shell Liquid is an important economic product of the plant, which is extracted from the shells of the raw nut and has various industrial uses like preparation of type writer rolls, drying enamels, water-proof coating for cement and brick flooring, manufacture of paints, varnishes and plastics. CNSL is a by- product of cashew industry, which is obtained from the shells. It is one of the few natural resins that is highly heat resistant and is used in the paint, automobile and foundry industry. Some of the advantages of CNSL based polymers are that it has improved flexibility, termite and insect resistance and anti microbial property.

6.3.1 Extraction Process

The sample units in Odisha and Tamil Nadu were following the expeller method for the extraction of oil from the shells. The extracted oil was filtered with the help of a filter press and then weighed and packed in barrels. Recovery of oil as percentage of 1 bag of shells (55 kg) was 19.38 per cent in Odisha and 21.8 per cent (Tamil Nadu) and the remaining were shell cakes. Expeller method being used was considered to be better than other methods like hot oil bath, kiln method, etc.

6.3.2 Economics of CNSL

Investment cost included costs on machinery like oil expeller, electric motor, accessories, weighing scales, storage tanks, etc. and building.

Table 14. Economics of CNSL Units* (Rs./kg)

Particulars	Odisha	Tamil Nadu
Shell cost	2.00	3.10
Operational cost^	2.30	3.87
Selling price of CNSL oil(Rs.1/hr)	15	13
Shelling price of shell cakes	0.60	2.30
Gross income	3.78	4.51
Net Income	1.48	0.64
FRR (%)	>50	15.26

* Data not estimated for Kerala

Cost on raw materials, electricity, packing material, wages, maintenance, etc.

Cashew nut shell, the main raw material for CNSL plants, constituted the maximum of total operational cost (88% in Odisha and 80% in Tamil Nadu) and the remaining costs were incurred on wages, electricity, maintenance, packing, etc. Cashew shells crushed/ year was 48,00,000 kg in Odisha and 4,95,000 kg (Tamil Nadu). The average operational days were 300 and 250 in Odisha and Tamil Nadu respectively.

The sample units were operating on a single shift of 8 hours a day and the average operational days for the units in Odisha and Tamil Nadu were 300 and 250 respectively. Full utilization of capacity of the unit was not realized in Odisha as CNSL processing was new to the region and the scarcity of raw materials also affected the functioning of the unit.

The net income derived was Rs. 1.48 and Rs. 0.64 per kg of cashew. The viability of the sample CNSL units in Odisha and Tamil Nadu was estimated at (>50 per cent) and 15.26 per cent respectively, which should encourage banks to finance such units and the entrepreneurs to undertake the activity on a commercial scale (Table 14).

6.3.3 Employment in CNSL Units

CNSL processing was being done more mechanically due to its irritant effect and the extraction as well as processing involves high temperature, boiling, drying, filtering, etc. The sample units were found to work on a single shift basis for 25 days in a month. The sample units were processing 900 bags of shells per month. Labour (skilled and unskilled) employed was 5 per day in Tamil Nadu and the same was 13 in Odisha.

6.3.4 Potential for CNSL and Cashew Processing

Though the country processes more than 1 million of raw nuts annually, the production of CNSL is limited. Cashew processing is a potentially lucrative activity that can be exploited by small-scale processors as it was found to be financially viable. Further, processing activity generated employment opportunities for rural women, especially for the weaker sections of the society.

6.4 Nutritive Value Per 100 g of Fresh Cashew Apple

Table 15. Nutritive value of cashew apple per 100 g

Principle	Nutrient Value per 100 g
Moisture	84.4 - 88.7 g
Protein	0.101- 0.162 g
Fat	0.05 - 0.50 g
Carbohydrates	9.08 - 9.75 g
Fiber	0.4 - 1.0 g
Ash	0.19 - 0.34 g
Calcium	0.9 - 5.4 mg
Phosphorus	6.1- 21.4 mg
Iron	0.19 - 0.71 mg
Carotene	0.03 - 0.742 mg
Thiamine	0.023 - 0.03 mg
Riboflavin	0.13 - 0.4 mg
Niacin	0.13 - 0.539 mg
Ascorbic Acid	146.6 - 372.0 mg

6.5 Health Benefits of the Cashew Fruit

6.5.1 Lower the risk of heart attack

Vitamin C in cashew fruit can get rid of atherosclerosis in the heart. Atherosclerosis is a state in which heart arteries thicken to increase heart attack due to blood so there smoothly. Cashew fruit consumption can be healthy for our heart.

6.5.2 Control the blood pressure

High blood pressure is a disease that is risky and can lead to sudden death in people who are suffering. Cashew fruit consumption by means of regularly can reduce blood pressure as well as heart and kidney damage.

6.5.3 Cure for asthma

People with asthma generally have a less amount of vitamin C in their bodies. Therefore, consumption of cashew fruit daily can reduce shortness of breath.

6.5.4 Act against Cancer

Many studies have shown that vitamin C may fight cancer in partial step. Not only that, the vitamin C in the fruit of the cashew can also minimize the effects of chemotherapy as nausea.

6.5.5 For Healthy eyes

Vitamin A in the fruit of the cashew act in the establishment of sight to the human senses. This vitamin may help convert the molecular signs of light received by the retina to be a projection of an image in our brains. Compounds acting foremost in this sort of thing is retinol. Together with rhodopsin, retinol compounds would create complex pigments are sensitive to the light beam to transmit sign to the brain. Regular consumption of cashew will help to nourish the organs of sight.

6.5.6 Prevent free radicals

In the cashew fruit there is a compound of beta carotene, a form of vitamin A called compound with a flurry of antioxidant that can prevent free radicals. Free radical compounds, many coming from oxidation reactions in the body or of pollution into the environment of the body.

6.5.7 Control anaemia

Cashew fruit has an iron content which is extremely useful to avoid anaemia. Cashew consumption in a way is able to control regular anaemia. Iron has the advantage that it is a major, which is to carry oxygen from the lungs to the tissues and transports electrons in the system of power formation in the cell.

(*Source:* http://www.vegetafruit.com/2015/12/health-benefits-of-cashew-fruit.html)

❑❑❑

Chapter 7

Marketing of Cashew

7.1 Marketing

Marketing in respect of cashew involved several players and channels. Marketing begins from the sale of raw cashew nuts by farmers and reaches the level of exporters/retailers for selling of processed and graded kernels to the ultimate consumers. The sample cashew growers sold a major portion of the produce to local traders, who in turn supplied the nuts to large traders and processing units located in Kollam (Kerala), Cuddalore (Tamil Nadu), Mangalore (Karnataka), etc. There are several entities in the marketing channels that get good share in the total spread between the producer and consumer. This Section analyses in detail the marketing aspects of raw nuts and kernels in the study area.

7.1.1 Marketing of Raw Cashew nuts

Marketing of raw cashew nuts in India has not yet been organized in a systematic manner except in Goa where co-operative marketing societies have a major stake in raw nuts trade. These co-operatives, where the producers were the major stakeholders acted as intermediary between the producers and the processors. The society had collection centres located in the production areas and procured cashew nuts from the growers. The sales price was fixed at Rs. 1.50 per kg above the procurement price and the processors had to lift the produce and bear the transportation cost from the society/collection centres. There was another co-operative set up, which directly procured raw nuts from producers and also had a processing unit on lease. Through this mode, the supply chain was further shortened and was beneficial both to producers and processors (Technical Digest, NABARD, 2007).

7.1.2 Marketing of Raw nuts through Regulated Markets

There was no regulated market for raw cashew nuts in Kerala and Odisha. Due to the absence of regulated markets, the farmers were forced to sell the raw nuts at prices determined by the local traders, who took a margin ranging from Re. 1 (Kerala & Tamil Nadu) to Rs. 2 (Kerala) per kg of raw nuts.

Even with the existence of regulated market for cashew in Panruti (Cuddalore district), raw nuts were sold by the farmers to the processors as well as commission agents, who visited the villages and collected the raw nuts from the farmers. Payment of cess and taxes in regulated markets deterred the producers from resorting to regulated markets.

7.2 Seasonality of Cashew nut Prices

Prices of cashew are linked to seasonality pattern like most other agricultural crops. The domestic prices of the crop are linked to new crop supply seasonality pattern at other origins, which supply raw nuts to India. As regards seasonal production, the harvest period in a growing region is quite short. Since the nuts can easily be dried and stored for at least a year, the processing industry is not very sensitive to finding continuous supplies. The relevance of seasonality is mostly to the anticipation of availability and therefore pricing of raw nuts. The peak seasons of output in different regions are approximately:

India, Vietnam, West Africa: March - June

Brazil: July - February

East Africa: October - December

The cashew nut arrivals of India, Vietnam, Côte d'Ivoire, Nigeria and Ghana coincide with one another, whereas the arrivals of Brazil, Indonesia and other African countries like Tanzania, Benin, Mozambique, Kenya coincide. Senegal and Guinea Bissau supply cashew nuts to the World during July and August. The seasonal index of imported raw nut prices in India has shown that the prices peak during September and October. The price of the locally produced cashew nuts is also influenced by the price and availability of imported nuts. When there is large inflow of imported cashew nuts, which are available at cheaper prices, the demand for locally produced nuts decline, thereby bringing down the prices.

7.3 Marketing Channels for Cashew

The prominent marketing channel prevalent in the sample districts is depicted below:

A case study of marketing dynamics of cashew in Cuddalore district revealed that ninety per cent of the processed kernels were sold through the export-marketing channel. In the export market, some of the other channels were similar to the above except that some of the exporters also owned processing units and the processing units/ exporters were also directly procuring raw nuts from the farmers.

Marketing of graded kernels by small cashew processors in Tamil Nadu were arranged through tie up with local export houses and traders; thus reducing their marketing risks.

In the domestic market, the price spread has been worked out for the prevalent marketing channel depicted in Table 16 below:

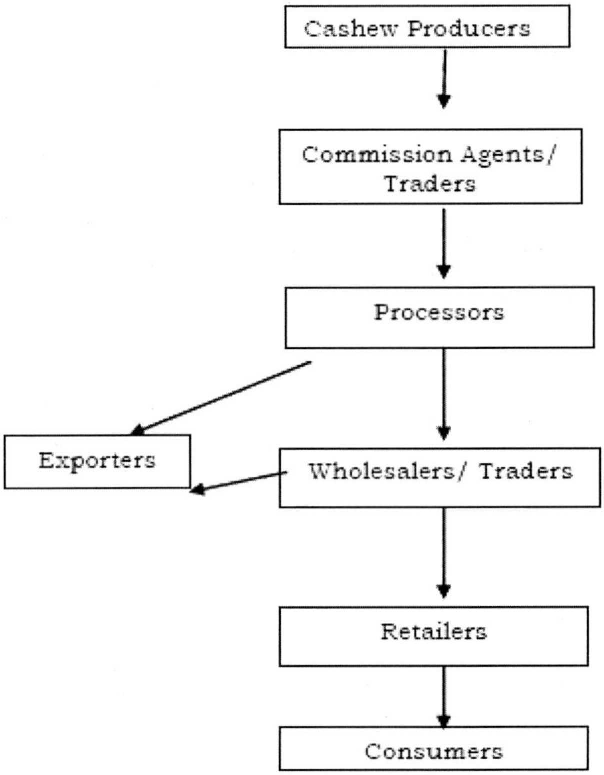

Fig. 16. Marketing Channel for Cashew

Despite the length of channel, the share of producer was 53.54 per cent in consumer Rupee and the price spread was Rs. 2342.75 per bag (80 kg) of raw nuts in Tamil Nadu. In the other marketing channel (Farmers→ Processing units→ Wholesalers→ Retailers→ Consumers), which was not widely prevalent, the share of the commission agents was only reduced. The commission agents charged Rs. 25 per bag (80 kg) of raw nuts as commission from the processors and acted as a link between the farmers and processors. Wholesalers purchased desired graded kernels from the processors and packed in different sizes of 1 kg, 500gm, 250gm, etc. and also in tins. Another category of wholesalers purchased the ungraded kernels (mixed grades) from the processors and graded and packed the kernels. Some wholesalers gave their own brand names to the pack.

The share of processors and wholesalers in the consumer rupee was 8.04 per cent and 15.52 per cent, respectively. No value addition was reported by the sample retailers but their share in consumer Rupee was 6.47 per cent (Table 16).

Table 16. Price Spread in Cashew Supply Chain

Particulars	Amount (Rs./bag*)	Share in consumers price (%)
Producers' Selling Price	2700	53.54
Commission charges	25	0.50
Processors' Margin	405.34	8.04
Wholesalers' Margin	782.80	15.52
Retailers' Margin	326.09	6.47
Miscellaneous charges (packing, handling, etc.)	803.52	15.93
Price paid by Consumer	5042.75	100

*1 bag = 80 kg of raw nuts; 23 kg of kernels were processed from 1 bag of raw nuts

Traders/wholesalers controlled cashew markets for both raw cashew as well as kernels. The cashew growers did not have any control over the market due to the absence of coordination and integration among themselves. As there were a number of intermediaries operating in the field between the primary producer and the processing unit, the different costs and margins in the total spread between the producer and the processing unit are quite significant and the producers' share in the price paid by the processing units is generally low.

7.4 Marketing Channel for CNSL

CNSL oil, extracted from cashew shells is widely used by the resin units in the fields of friction materials, adhesives, etc. The prevalent marketing channel for CNSL was:

Farmers → Commission agents → Processing units → CNSL Plant → Resin Plant → Paint Industry

The sample units in Tamil Nadu were selling the manufactured CNSL oil @ Rs. 13 per litre directly to the resin units in Chennai and Hyderabad and the same was supplied to the traders in Kolkata @ Rs.15 per litre by the sample units in Odisha. Cashew shell cake is another by-product of the CNSL unit and was sold as fuel @ Rs. 0.60 and Rs. 2.30 per kg in Odisha & Tamil Nadu respectively. In Tamil Nadu, the operational cost for processing CNSL worked out to Rs. 212.97 per bag (55 kg shells). The selling price was Rs. 248 for the output that included 12 kg filtered oil and 40 kg shell cakes. Thus, a margin of Rs. 35.03 was available for CNSL plant in Tamil Nadu from 1 bag (80 kg) of raw nuts. The sample CNSL units were selling oil to the resin units who incurred a cost of Rs. 30 per litre and sold to the paint industry @ Rs. 35 per litre.

7.5 Futures Trading in Cashew

Cashew futures are exchange traded contractual obligations to make or accept delivery of a specified quantity and quality of cashew during a specified time in the future at a price agreed upon at the time the commitment is made. At present, futures are available in cashew at National Commodity and Derivatives Exchange Limited (NCDEX) and MCX. MCX was the first commodity exchange in the world to start futures trading in cashew. NCDEX has launched a cashew futures contract in Kollam, Kerala since March 2005. The price quote is on net basis and net weight of each carton is 22.68 kg. Trading is done for white wholes, with a count of 300-320 nuts per 454 gm. It is stipulated that the kernels should be free from infestation, insect damage, mould rancidity, adhering testa and extraneous matter. The delivery centres are located at Kollam, with an additional delivery centre at Mangalore.

Although, the commodity futures help the exporters in hedging against price fluctuations as they can sell the commodity at a price decided months before the actual transaction, thus ironing out any fluctuation in prices that happen subsequently; the sample processors/exporters were not trading on the future exchange.

7.6 Cashew Export Promotion Council (CEPC)

The Cashew Export Promotion Council of India set up in 1955 with the processors and exporters as members is actively engaged in the export promotion of cashew kernels and cashew nut shell liquid. The Council undertakes publicity abroad to highlight the quality of Indian cashew and also participates in international fairs and exhibitions. High quality is a major criterion for success in the world market. Quality control is administered through the Cashew Export Promotion Council. It also helps in settling complaints amicably in matters related to export/import. The Cashew Export Promotion Council of India is to implement a scheme titled 'Modernisation & Diversification Scheme' with the following four components during the eleventh plan (CEPC):

- Process Upgradation/Modernization of Cashew Processing units
- Setting up facilities for Value Addition/Consumer Packing of Cashew Kernels
- Adoption of New Packaging Systems
- Implementation of Internationally Accepted Quality Systems (ISO, organic, etc).

The grant assistance would be 25 per cent of the total eligible investment, of which civil works should not constitute more than 50 per cent subject to a maximum total grant of Rs. 50 lakh per exporter for all components put together. Exporters who are members of the Cashew Export Promotion Council of India are eligible for assistance under this scheme. Some of the items eligible for grant assistance are indicated below:

- Process Upgradation/ Modernisation of Cashew processing units include boiler, cutting machines & accessories, oil expeller for CNSL extraction, Hot Air Generator & Accessories Trolleys, steel/aluminium trays, etc.
- Setting up facilities for Value Addition/Consumer Packing of Cashew Kernels-Cashew kernels roasting machinery, salting/coating/flavouring machinery, heating and cooling equipment for production processing, consumer packaging machines, printing machines for consumer packs, labeling machines, etc.

- Adoption of new packaging system- Flexible packaging machine, accessories mould & former filling system, Air Compressor, conveyor vibratory table with vibratory motor, electronic weighing machine, grading machines, automatic packing systems, pouch forming machines, etc.
- Implementation of internationally accepted quality systems- ISO/ organic consultancy charges, certification charges, air curtain, vacuum cleaner, building and related civil works, laboratory equipments, etc.

7.7 Specifications for Cashew Exports

Government of India has laid down standard specifications for the cashew kernels for export under the Export Control and Inspection Act, 1963. The Act prescribes 33 different grades of cashew kernels but only 26 grades are commercially available and exported. Cashew kernels are selected on the basis of the number per unit weight, generally per pound of weight. They are also graded into white / scorched wholes, pieces, splits, butts, etc. depending on the shape, size & colour of the kernel. White or ivory kernels are preferred over brown ones. There is a maximum permitted moisture level of 5 per cent (both for raw cashews and cashew kernels) and the product must be free from insects, mould and extraneous materials.

7.7.1 Quality Control and Pre - shipment Inspection

The export of cashew kernels from India is subject to voluntary quality control and pre- shipment inspection. Under inspection, it is ensured that the product is processed and packed as per the standards prescribed.

7.7.2 Packing and Standard weight

The standard net weight of cashew kernels packed in tins is 11.34 kg or 25 lbs and the net weight of each carton containing two such tins is 22.68 kg or 50 lbs.

7.7.3 Shipping Specifications

Cashew shipments from India are mainly in 20 ft containers, which have 650-700 cartons. Freight is charged per cubic metre on volume basis.

7.7.4 Contract Terms

The contracts are normally made on FOB or C&F basis, cost of insurance to be borne by the buyer, export dealings are finalized on the basis of opening of 100 per cent irrevocable LC by the buyer in favour of the shipper. Besides, fumigation certificate, certificate of origin, etc. are also required.

7.8 Agri Export Zone for Cashew in Cuddalore District (Tamil Nadu)

To set up a Cashew Export Zone at Panruti in Cuddalore district, a Memorandum of Understanding was signed between GoTN and APEDA (Agriculture and Processed Food Products Export Development Authority) at a project cost of Rs.17 crore in April 2005. M/s Sattva Agro Export Pvt. Ltd. has been selected as the anchor promoter for cashew Agri Export Zone through private participation. The work for AEZ commenced in 2006-07 with the construction of infrastructure facilities. Modern nursery has been established for providing quality seedlings to the farmers and the construction of processing unit and pack house has been completed (Agriculture Policy Note, 2009-10, Government of Tamil Nadu, 2009-10).

7.9 Trend in Indian Cashew Trade Flow

Now India stands first in import of raw nuts with nearly 30% of total import in the world. At present, 3797 processing units are functioning in India under organized and unorganized sector with a processing capacity to process 16.23 lakh MT raw nuts per year. The present domestic production raw nut is 7.36 lakhs tones. The industry, therefore, imports more than percent of raw nuts through imports from other countries, especially from Africa to meet the demand. The major source of raw cashew nut import is from Ivory Coast, Benin, Tanzania, Guinea Bissau, Ghana and Indonesia. The increased dependency on imports is fraught with danger as the exporting countries could start processing their own raw nuts. Hence, efforts needed to achieve self-sufficiency in raw nut production. (Jnanadevan, 2016).

The available data shown in the table-17 indicates that the quantity of raw nut processed in India increasing every year. The global market for cashew kernels is growing steadily at an average compound growth rate of 7% annually. In 1998-99, the total raw nuts processed in India was 6.41 lakh MT that increased to 15.72 lakh MT in 2013-14. When the import of raw cashew nut was increasing, There was a steady increase in domestic production also. However, it was not sufficient to meet the growing demand of cashew processing industry. Hence, Indian cashew industry's dependence on other producing countries for raw cashew nut appears to have become a continuous phenomenon, as indigenous supply has not been able to keep pace with the processing industry's ever growing requirement. The trend in export was found directly correlated with import of raw nuts during the last fifteen years except for last few years. From 1998-99 onwards, there was a steady increase in export of kernel and raw nut import except during 2008-09 to 2010-11 and 2012-13. The export was decreased marginally by 5% in volume in the year 2010-11 and 23.5% in quantity in 2013. However, there is an in-creasing trend again noticed during the year 2013-14. The strong domestic demand and shortage of raw nuts had lead decline in India's cashew kernel export in these years. The global economic recession also affect the cashew kernel export adversely.

The annual increase in cashew kernel export was massive and all time highest, *ie,* 24% in quantity and 55% in value during 2011-12 compared to 2010-11. The quantity of raw cashew imported was also showed significant increase during the above period. The import increased from 5.29 lakh MT in 2010-11 to 8.92 lakh MT in 2012-13. But in 2012-13 the quantity of cashew kernel exported has decreased by 23.5% in 2012-13 compared to the previous year. The export has decreased from the all time high of 1.308 lakh MT to 1.00 lakh MT in the year 2012-13. The export then increased in 2013-14 by 11%. The year 2012-13 has been a tough year for Indian cashew Industry. The demand and price peaked at the beginning of the year 2012 and fell throughout the year which made it very difficult year for the exporters. The ever recorded Indian rupee depreciation has made imports costlier in turn affecting the export of cashew kernel.

The contribution of domestic production to the total raw nuts available for processing shows a decreasing trend. From 1998-99, the percentage of domestic production contribution seen steadily decreasing from 72% to 29%. Similarly, when the export was increasing there was steady decrease in the contribution of domestic production in export. Exports of cashew kernels have also been showing a steady increase, from 75793 tonnes 1998-99 to 130869 MT in 2011. The export earnings increased at an average of 11.4% per

Table 17. Export, import and indigenous production of cashew in India

Year	Export (CK in 000 MT)	Import (R.N in 000 MT)	ValueRS. In Crs.	Indigenous production utilized (R.N in 000 MT)	Total qty. R.N available	% in export	% of Domestic consumption
1998-1999	75.593	181.009	680.240	460.000	641.009	51	49
1999-2000	92.461	200.584	953.350	520.00	720.584	56	44
2000-2001	89.155	249.318	960.800	450.000	699.318	55	45
2001-2002	97.550	356.566	960.010	470.000	826.566	51	49
2002-2003	104.137	400.199	1236.570	500.000	900.199	50	50
2003-2004	100.828	452.898	1400.900	535.000	987.898	44	54
2004-2005	126.667	578.884	2183.240	544.000	1122.884	49	51
2005-2006	114.143	565.400	2162.950	573.000	1138.400	43	57
2006-2007	118.540	592.604	1811.620	620.000	1212.604	42	58
2007-2008	114.340	605.970	1746.800	665.000	1270.970	38	62
2008-2009	109.523	605.654	2632.410	695.000	1300.970	37	63
2009-2010	117.991	752.894	3037.350	613.000	1365.894	34	66
2010-2011	105.755	507.138	2499.75	653.000	1160.138	40	60
2011-2012	130.869	809.825	5338.640	692.000	1501.825	38	62
2012-2013	100.105	892.365	5331.740	728.474	1620.839	27	73
2013-2014	111.791	836.000	5058.730	736.500	1572.500	29	71
2014-2015	117.368	932.289	6570.93	725.420	1657.709	28	72

Source: DCCD, Kochi

annum during the last 15 years, ie from Rs. 1609.90 crores in 1988-89 to Rs. 4067.2 crores in 2012-13. Imports of raw nuts have gone up from 181009 MT valued at Rs. 680.26 crores in 1989-90 to 8,36,060 MT valued at Rs. 5058.73 crores in 2013-14. The rate of growth in raw nut import was at average of 26% per annum and the average rate of increase in the money spends for import of RCN was at 45% during the period. Indigenous availability of raw nuts was also increased from 4.6 lakh MT in 1998-99 increased to 7.36 lakh MT in 2013-14. (Jnanadevan, 2016).

❑❑❑

Chapter 8

Strategies for Cashew Development

8.1 Issues in Cashew Cultivation

8.1.1 Low Productivity

Large area under cashew is covered with non- descript genetically inferior seedling progenies. Compared to other plantation crops, cashew is still confined mostly to marginal and poor fertile lands and is considered as a wasteland crop. Moreover, cashew has been considered as 'maintenance free' crop and the recommended package of practices are not followed. All these factors lead to low yield.

8.1.2 Senile Plantations

About 40 per cent of the existing cashew areas have become senile resulting in uneconomic production. These plantations generated from indiscreet seedlings are least responsive to the technological packages and have adversely affected the productivity and competitiveness of cashew.

8.1.3 Poor quality of Raw Cashew nuts

Incidence of pests and diseases like tea mosquito, cashew stem and root borer, unsatisfactory drying of raw nuts and inadequate storage of dried nuts have resulted into poor quality of raw nuts produced. The yield loss due to tea mosquito bug infestation ranged between 30 and 50 per cent in different years, while the stem and root borer infestation in neglected plantations ranged around 8 to 10 per cent (NRCC, Puttur). Poor quality of raw nuts in turn leads to inferior quality of processed kernels.

8.1.4 Unorganized Cultivation

Cashew is predominantly a small holder crop in India as 70 per cent of cashew is grown by small farmers. The remaining 30 per cent are grown under re-forestation programmes. Hence, it is not cultivated as an organized plantation crop.

8.1.5 Infrastructural Facilities

Godowns for storing raw nuts are inadequate. Deterioration rate is low for properly dried nuts. Drying the nuts immediately after harvesting is essential to preserve their quality and reduce moisture content. Russell (1969) considered whole nut moisture content of 9 per cent or below to be safe for storage. Sundrying of raw nuts is usually done and after drying they need to be stored and protected from rain and stored in local godown as soon as possible. Processing industries also require storage facilities for storing raw material for a year's production. Thus, godowns will facilitate buying operations and nuts may be marketed regularly.

8.1.6 Non-availability of Grafts

Since high yielding cashew grafts are not available in sufficient quantity during the planting season, the farmers are going for seedling planting subsequently results in lower yields. Government of India through Directorate of Cashew and Cocoa Development (DCCD) gives assistance to nurserymen to develop Regional Nurseries with increased cashew graft production. Farmers also encouraged going for new planting of grafts through state Department of Horticulture schemes.

8.2 Issues in Cashew Processing

8.2.1 Insufficient Domestic Raw nuts

As per the statistics of DCCD, the production of raw nuts in India during 2014-15 was 7.25 lakh tonnes as against the requirement of 16 lakh tonnes by the cashew processing industries. Thus, the availability of domestic raw nuts was only 45 per cent of the processing capacity. This has led to dependence on imported nuts. Cashew producing countries in Africa and South-East Asia, which were traditional suppliers of raw nuts to India are developing their cashew processing facilities to process and export kernels. Development of cashew processing in these countries will affect the availability of raw nuts for import into India.

8.2.2 Working Capital Requirements

Working capital is required for purchasing raw cashew nut inventory. Small processors without boilers had constraint in accessing working capital limits from the banks. Access to working capital would induce entrepreneurs to invest in cashew processing facilities.

8.2.3 Traditional Cashew Processing

In India, most of the processing takes place on a small scale and is highly labour intensive. The main advantage of manual conversion is low cost. However, mechanization to improve the recovery and quality of kernels, better extraction of CNSL have not taken place unlike other countries.

8.2.4 Inadequate Skilled labour for Processing

Shortage of power supply and inadequate skilled labour for processing activities like shelling, drying, etc. becomes a constraint for the processing units. Main difficulties in shelling cashew nuts are the irregular shape of the nut and the CNSL within the shell, which must not contaminate the kernel during the process. Drying of shelled kernels is necessary to avoid pest and fungal attack and to facilitate peeling. Kernels are dried to moisture content of 3 per cent and should not be excessively dried resulting in high breakage percentage.

8.3 Issues in Cashew Marketing

8.3.1 Competition from other Countries

Vietnam is the major competitor for India for cashew kernels. In order to promote domestic processing and exports, Vietnam Government imposed 15 per cent tax on export of raw nuts in 1995, which resulted in decrease of export of raw nuts. Brazil and Vietnam compete with India in purchasing raw nuts.

8.3.2 Declining Domestic Raw nut Prices

Raw nuts are imported mainly from African countries like Nigeria, etc. at a cheaper rate than that of local nuts. This has posed a problem for the farmers. During 2007-08, the price of raw nuts declined to Rs. 32.20/kg from Rs. 40.89 kg in 2005-06.

8.3.3 Poor Quality of Processed Kernels

The quality of the processed kernels is not good, as the small processors do not maintain hygienic conditions in their factories. This affects the marketability of kernels in the international market.

8.3.4 Competition from other nuts

One of the major factors that affect the consumption of cashew kernels in the world market is the competition from other nuts. The major importers in developed countries contract their requirements for the whole year based on the sales from previous years. Since cashew cultivation is not organized on a plantation scale in most of the producing countries, there is a fluctuation in the yield every year, which leads to wide price fluctuations of cashew kernels. On the other hand, other nuts like almond and pistachio are grown in large plantations and thus their prices are steady.

8.3.5 Intermediaries in Supply Chain

Traders and middlemen dominate the market for raw cashew nuts and kernels. Since procuring raw cashew is the largest component of the operating costs in cashew processing sector, a slight increase in cashew price adversely affects the entire economics of cashew processing. The individual farmers were in a disadvantageous position as they were forced to sell the produce at a price determined by the traders/leaseholders. Farmers did not use the regulated markets in Cuddalore, as taxes/cess had to be paid by the traders if it was sold through the Marketing Committee.

8.4 Drain on Foreign Exchange Reserves

During 2014-15, the country imported 9.40 lakh tonnes of raw cashew nuts to meet the requirements. This has considerable drain (Rs. 6571 crores) on the country's foreign exchange reserves.

8.5 Strategies for Cashew Development

- Technologies like use of vegetative propagated planting materials may be used for increasing the production and productivity of cashew. The present level of productivity is 860kg/ha, whereas the new varieties have a potential of 2000 kg/ha. Efforts need to be taken for replacement and new plantation with clones of these varieties. Trials with different density and spacing in cashew have proved that net profit realised from high tree density planting (384 trees/ha, 6.5 x 4m) was high for the first 10 years compared to normal tree density (156 trees/ha). Better management practices like pruning, top working for rejuvenating cashew trees, improved planting material, adequate disease and pest control, etc. are required to increase the yield.

- Senile plantations adversely affected the productivity and competitiveness of cashew. Production and productivity can be enhanced through a phased replanting programme. Farmers were hesitant to take up replantation in the study area due to the expected crop loss and negative returns during the initial years. Inter cropping may be practiced by cashew growers as it helps in obtaining returns from cashew. Crops and trees like groundnut, cowpea, tapioca, casuarinas, turmeric, black gram, etc. can be effectively grown as inter-crop during the initial 4-5 years. Strong extension activity and credit support is required to make the farmers rejuvenate old plantations as well as to practice intensive cultivation practices.
- Organic cashew offers new opportunities for the producers as they command price premium. With the support of the State Department of Horticulture in Tamil Nadu, the farmers had taken up partial organic cashew cultivation. But concerted efforts are required for promoting certified organic cashew. Though most of the cashew produced in India are organic as very less/no fertilizers are used in the plantations, they need to be certified as organic.
- There was no contract farming arrangements for cashew cultivation in the study area. This was mainly because cashew has not developed into organized plantation. It may become feasible by setting up of Cashew Export Zone at Panruti in Cuddalore district, for which a Memorandum of Understanding has been signed between GoTN and APEDA. Contract farming can evade middlemen between the farmers and the processors and ensure adequate prices to cashew farmers.
- Better transfer of technology may bridge the gap between the potential and actual yields of cashew. The farmers may be educated about the potential of cashew sector and encouraged to adopt advanced farming techniques so that they can earn maximum benefits.

- Cashew is the only major plantation crop that is not regulated by an autonomous board. Other plantation crops such as tea, coffee, cardamom and rubber all have efficient and autonomous boards and as a result have experienced much faster growth in productivity than cashew. Considering the importance of the crop, setting up of the Board may be made by the Commerce Ministry. A separate cashew development board may be set up so as to enhance the cultivation of raw nuts and increase the production and export of cashew kernels.
- Cashew apple preparations like jam, jelly, chutney, juice, syrup, etc. need to be popularized and commercially exploited, as it will increase the income of cashew cultivators and also enhance rural employment. Except in Goa, cashew apple is wasted in almost all the states in India. There is also a need to popularize the techniques of manufacturing the value added products from cashew apple among the SHGs.
- Scientific processing techniques to recover cashew shell liquid oil may be used in the processing of raw nuts.
- The role of middlemen in the market should be reduced and Government procurement system should be strengthened so as to motivate farmers to grow this crop on a sustainable basis and ensure a better price.
- Establishment of cashew clusters among the processors may facilitate the expansion of market linkage and improvement of quality of kernel. Cluster approach would also facilitate setting of other ancillary units like CNSL, units producing jam, pickles, etc. from cashew apples.
- Infrastructural facilities like godowns for storage of raw nuts may be constructed. If godowns are used, the factory will need small storage capacity and will facilitate buying operations of nuts. Development of infrastructure like nurseries and extension services can pursue the cashew growers to accept cashew as a commercial plantation crop.
- Institutional support and co-ordination among various developmental and promotional agencies such as DCCD, CEPC, Agricultural Universities, etc. would help in promoting cashew processing and marketing.

8.6. Good Agricultural Practices (GAP)

8.6.1 Introduction

Low productivity of cashew in India is mainly due to planting of seedlings and wrong choice of varieties and poor management of the plantation. Water, nutrition management and pest management is completely absent in most of the casherw growing areas. Following good agricultural practices not only improves the productivity, the quality of nut produced also improved. Gorakhsingh *et al.* (2016) explained the important aspects in Good Agriculural Practices in Cashew.

8.6.2 Site Selection, soil and climate, rainfall

Site selection should be done very carefully before going for cultivation. The nature of soil, climate prevailing in that area should be taken into account before going for large area commercial cultivation.

Cashew is a hardy crop and can be grown in most of the soil except waterlogged, saline and heavy clay soils. Well drained, lateritic, red soil or coastal sandy soils are suitable for cashew cultivation.

Coastal climate and elevation upto 700 m mean sea level is ideal for cashew cultivation. It is a drought hardy crop but will not tolerate frost. Ideal temperature for normal growth and development is 24-28 C. Cashew can tolerate temperature above 36 C for shorter period. But high temperature (39-42 C) during fruit setting may cause fruit drop.

Cashew is successfully grown in wide range of rainfall of 600 mm to 4500 mm per annum. Rainless period during flowering is ideal for fruit setting.

8.6.3 Selection of Variety and Nursery management

The varieties recommended for that locality or state from the reputed Research Stations should be grown. Under All India Co-ordinated Research Project on Cashew, ICAT institutes and State Agricultural Universities have released more than 34 varieties. Example. VRI-3 and VRI(cw)H1 for Tamil Nadu from Tamil Nadu Agricultural University.

The production of scion sticks, selection of scion, procuring, collection of scion sticks, raising of rootstocks, potting mixture, sowing of seeds and maintenance of rootstocks, softwood grafting ans management of grafts should be done as explained in the previous chapters.

8.6.4 Field preparation and planting

Recommended spacing for the varieties either normal spacing or high density planting may be followed. According to different planting systems, the number of trees per hectare may vary from 150 to 1250 trees per hectare. Spacing, planting, after care, canopy management, and soil and water conservation is explained in chapter III.

8.6.5 Integrated Nutrient Management

Depending upon the soil type, age of the crop and variety, the nutritional requirement of cashew varies from place to place. The number of plants per hectare, either normal planting or high density planting also decides the fertilizer requirement. The nutrition application through soil, foliar and fertigation is explained in chapter III.

8.6.6 Integrated Pest Management

Major pests like cashew stem and root borer, tea mosquito bug can cause yield reduction to the tune of 30-40 per cent. Wrong choice of chemicals may lead to emergence of minor sucking pests in to major pests. Hence, choice of chemicals should be done carefully. The chapter IV explains the management of pest and diseases of cashew.

8.6.7 Post Harvest Management

The harvested nuts are sun dried for 2-3 days to reduce the moisture from 25% to below 9%. It is very essential to dry cashew nuts after harvest to prevent spoilage during storage. With proper drying, the kernels retain their quality, especially the flavour. Polythene lined gunny bags (300 gauge) may be used to store the nuts.

The cashew apple which is a byproduct of cashew harvested is normally wasted or put in manure pits. Hand harvested cashew apples can be successfully converted in to value added products like syrup, squash and various products as explained in chapter VI.

❏❏❏

Chapter 9

References

Abeysinghe, D.C., Sangakkara, U.R. and Jayasekera, S.J.B.A. 2003. Inter-cropping of young cashew (*Anacardium occidentale* L.) and its effects on crop productivity and land utilization. *Trop. Agric. Res.* 15: 10-19.

Adebola, P.O. 1999. Effect of nut weight on germination and seedling vigour in cashew. *Nigerian Journal of Tree Crop Research,* 3(2): 71-79.

Adebola, P.O., Esan, E.B. and Famaye, A.O. 1999. Effect of nut weight on germination and seedling vigour in cashew. *Nigerian Journal of Tree Crop Research,* 2(1): 6-11.

Adeyemi, A. A. (1989). Cultural weed control in cashew plantations: Use of inter-crops to reduce weed incidence in cashew plots. Proceedings: Integrated Pest Management in Tropical and Subtropical Cropping Systems '89, Volume 3. Bad Durkheim, Germany, February 8-15: 827-42.

Adeyemi, A.A. 1998. Effects of inter-cropping on weed incidence in cashew (*Anacardium occidentale*) plantations. *Nigerian J. Tree Crop Res.,* 2: 83-94

Aguiar, M de J.N. and Costa, C.A.R. 2002. Exigências Climáticas. p. 21-23. *In*: L. de M. Barros (ed.) Caju. Produção: Aspectos técnicos. Embrapa Agroindústria *Tropical. Fortaleza.* (Frutas do Brasil, 30).

Akinwale,T.O. 2000. Cashew apple juice: Its use in fortifying the nutritional quality of some tropical fruits. Euro Food Res. Technol., 211: 205-207.

Albert Alex, V. and Krishnasamy, V.2004. Evaluation of organic sources of nutrients on growth and yield parameters of tomato, Master thesis, Department of soil science and agricultural chemistry, TNAU, Coimbatore-3.

Aliyu, O. M. and Awopetu, J.A. 2005. *In-vitro* regeneration of hybrid plantlets of cashew (*Anacardium occidentale* l) through embryo culture. *African Journal of Biotechnology,* 4(6): 548-553.

Aliyu, O.M. and Awopetu, J.A. 2007. Assessment of genetic diversity in three populations of cashew (*Anacardium occidentale* L.) using protein-isoenzyme electrophoretic analysis. *Genet. Resour. Crop Evol.*, 54: 1489-1497.

Ambethgar, V.2010. Field assessment of fungal and indecticidal delivery methods on cashew stem and root borer, *Plocaederus ferrugineus* L. (Cerambycidae: Coleoptera). *J. Biopesticides*, 3(1): 121-125.

Amoah, F.M. 2005. The germination and early growth of cashew (*Anacardium occidentale* L.). *Tropical Science*, 45(4): 149-152.

Ananthakrishnan, G. 1999. In vitro plant regeneration from embryo axis of cashewnut. Plant Tissue Culture & Biotechnology Emerging Trends: 123-128 (Proceedings of the symposium held at Hyderbad in 1997).

Ananthakrishnan, G., Ravikumar, R and Girija, S. 2002. In-vitro adventitious shoot formation from cotyledon explants of cashew. *Scientia Horticulturae*, 93(3/4): 343-355.

Ananthakrishnan, G., Ravikumar, R. and Girija, S. 2004. Selection of efficient arbuscular mycorrhizal fungi in the rhizosphere of cashew and their application in the cashew nursery. *Sci. Hort.*, 100: 369-375.

Aneesarani, M.S. 2007. Foliar spray for improving fruit set and nut yield in cashew variety VRI 3. In: National seminar on Research, development and marketing of cashew, souvenir and extended summaries, ICAR Research complex for Goa, 20-21st November, 2007.

Aneesarani, M.S., Jeeva, S., Allathanbi, G.N. 2004. Bio-fertilizer studies on growth, proliferation of shoots and roots of cashew rootstocks to be used for grafting. Abstracts, First Indian Horticulture Congress, New Delhi.

Aneesarani, M.S and Keisar Lourdusamy, D. 2016. Experimental results of ultra high density planting of cashew (In Tamil) In: Scientific Agriculture in Tamil-2 Eds. K. Ramasamy, P. Pandiarajan, M. Jawaharlal, S.Sekar and M.Manimekalai. ISBN: 81-902877-3-7 Publ. by AC&RI, TNAU, Trichirappalli.

Aneesarani, M.S, Kumar, N. and Marimuthu, R. 2011. Evolving cashew F1 hybrids suitable for high density planting system. *Indian J. Hort.*, 68(2): 152-155.

Aneesarani, M.S. and Jeeva, S. 2010. Bio-fertilizers for enhancing growth of cashew rootstocks. *Adv. Plant Sci.*, 23(10): 31-34.

Aneesarani, M.S. and Jeeva, S. 2010. Heterosis for yield and nut weight in cashew. *Adv. Plant Sci.*, 23(1): 213-214.

Aneesarani, M.S., Jeeva, S. and Vaidyanathan, R. 2011. Effect of pruning and foliar spray on growth and yield of cashew variety VRI 3. In: 1st International symposium on cashew nut, held at AC & RI , TNAU, Madurai, 09-12 December 2011.

Aneesarani, M.S., Jeeva, S. and Vaidyanathan, R.2011.Profitable intercrops for cashew. In:International Symposium on Cashew Nut Souvenir and Abstracts,9-12 December, Agricultural College and Research Institute, TNAU, Madurai, TN, India, pp. 110.

Anil, S.R. and Thimmappaiah. 2005. Somatic embryogenesis from nucellar callus of cashew. *Journal of Horticultural Science and Biotechnolgy*, 80(3): 327-337.

Anil, S.R., Thimmappaiah and Rohini Iyer, 2000. Vesicular arbuscular mycorrhizal association in cashew. Recent Advances In Plantation Crops Research 13: 390-392 (Proceedings of PLACROSYM XIII held at Coimbatore, 1998).

Anonymous. 2011. High density planting system in cashew. In: New crop implements and management technologies, Ed: Directorate of Research, Tamil Nadu Agricultural University, Coimbatore-641 003.

Archak, S., Gaikwad, A.B., Gautam, D., Rao, E.V.V.B., Swamy, K.R.M. and Karihaloo, J.L. 2003. Comparative assessment of DNA fingerprinting techniques (RAPD, ISSR and AFLP) for genetic analysis of cashew (*Anacardium occidentale* L.) accessions of India. *Genome*, 46: 362-369.

Avilán, R.L. 1971. Efeitos e sintomas deficiências de macronutrientes no crescimento e na composição do cajueiro (*Anacardium occidentale* L.) Dissertação (Mestrado) - Universidade de São Paulo, Piracicaba.

Badhe, V.T. and Mayar, S.S. 2004. Influence of different conservation measures on runoff, soil and nutrient loss under cashew nut in lateritic soils of south Konkan region. *Ind. J. Soil Conser.*, 32: 143-147.

Badrinath, M.S., Udhir, K. and Chikkaramappa, T. 1997. Soil fertility evaluation for cashew cultivation in coastal Karnataka soils. *The Cashew*, 11: 22-23.

Balasimha, D. and Yadukumar, N.1993. Effect of plant density on photosynthesis in cashew. *J. Plant Physiology*, 36: 5-7.

Balasubramanian, D. 1998. Cashew nut processing industries in India, Harvest and Post-harvest Technology of Plantation Crops, (eds. Bosco, S.J.D., Sairam, C.V., Muralidharan, K. and Amaranath, C.H.). CPCRI, Kasargod, pp. 102-110.

Barros, L. de M. 1995. Botânica, origem e distribuição geográfica. p. 55–71.*In*: J.P.P de Araújo, and V.V. da Silva (ed.) Cajucultura: Modernas técnicas de produção. Embrapa-CNPAT. Fortaleza. (Frutas do Brasil, 30).

Barros, L. de M., Paiva, J.R. and Cavalcant. J.J.V. 1998. Cajueiro anão precoce. Melhoramento genético: *Estratégias e perspectivas. Biotecnologia*. 2: 18-21.

Basu Choudhuri, J.C. and Misra, M.P. 1962. Preliminary investigation on the insect pests of cashew plantations in Kerala. *Indian Forester,* 88: 516-522.

Beevi, S.P. and Mahapatro,G.K.2007. A new field screening methodology for cashew genotypes against tea mosquito bug. *Helopeltis antonii* Sign. *J. Plantation Crops* 35(3): 139-145.

Bhat, P.S. and Raviprasad, T.N. 2008. Pheromones of tea mosquito bug *Helopeltis antonii* Signoret, Focus on Research, *Cashew News,* January-June 2008, National Research Centre for Cashew, Puttur, Karnataka, pp 2-3.

Bhat, M.G., Kumara, P.M. and Thimmappaiah. 1998. Pollination technique in cashew. *The Cashew*. 12: 21-26.

Bhat, M.G., Yadukumar, N. and Gangadhara Nayak, M. 2007. Cashew research achievements, technologies developed and research and development strategies. **In**: *Souvenir on National Seminar on Research, Development and Marketing of Cashew,* Nov. 20-21, 2007, ICAR Research Complex for Goa, Ela, Old Goa, India.

Bhattacharya, A. 2004. Flower visitors and Fruit set of *Anacardium occidentale. Annales-Botanici- Fennici.* 41(6): 385-392. (Palynology Laboratory, National Botanical Research Institute, Lucknow-226 001, India).

Bhattacharya, A. 2005. Age-dependent pollen abortion in cashew. *Current Science.* 88(7): 1169-1171. (Palynology Laboratory, National Botanical Research Institute, Lucknow-226 001, India).

Busov, B.V., Meilan, R. and Strauss, S.H. 2003. A short route to gene discovery in trees: Control of stature through gibberellin catabolism. *ISB New Report*. Sept. pp. 1-4.

Chacko, E.K. 1994. Cashew hybridization during 1988–92 (CSH–36 H) and evaluation of progenies planted at four test sites in Northern Australia. Working papers of the seventh Annual cashew research and development workshop, May 17, 1994, Cairns, North Queensland. Pp. 1-10.

Chakraborti, S. and Chakraborti, A. 2009. Promoter factors for cashew stem and root borer incidence. *J. Applied Zoological Researches*. 20(1): 5-7

Crisóstomo, L.A. 1991. Avaliação da fertilidade em dez unidades de solo cultivadas com cajueiro nos estados de Piauí, Ceará e Rio Grande do Norte. Tese (Professor Titular). Universidade Federal do Ceará, Fortaleza.

Crisóstomo, L.A., Santos, F.J. de S., Oliveira, V.H., van Raij, B., Bernardi, A.C. de C., Silva, C.A. and Soares, I. 2003. Cultivo do cajueiro anão precoce: Aspectos fitotécnicos com ênfase na adubação e irrigação. Embrapa Agroindústria Tropical. Fortaleza.

Croxford, A.E., Robson, M. and Wilkinson, M.J. 2005. Characterization and PCR multiplexing of polymorphic microsatellite loci in cashew (*Anacardium occidentale* L.) and their cross-species utilization. *Mol. Ecol. Notes*. 6: 249-251.

Deshmukh, M.T., Sawke, D.P., Borude, S.G., Hurni, H. and Tato, K. 1992. Effects of platform bench terraces on the growth and yield of mango and cashew grafts. Erosion, conservation, and small scale farming report. Department of Agricultural Engineering, Konkan Agriculture University, Ratnatgiri, India. pp. 477-482.

Deshpande, S.B., Haldanker, P.M., Patil, B.P., Athalye, N.D., Jambhale, N.D. and Rao, E.E.V.V. 2003. Evaluation of pollen parents as polliniser varieties in cashewnut. *The Cashew* 17(1): 23-29.

Falade, J.A. 1984. Variability in soils and cashew tree size. *Journal of Plantation Crops*, 12: 30-37.

Fragoso, H.A. 1999. Macronutrient removal by both cashew nut and apple of two dwarf cashew clones. *Revista Brasileira De Ciencia Do Solo*: 23: 603-60.

Frota, P.C.E. and Parente, J.I.G. 1995. Clima e fenologia. p. 43-54. *In*: J.P.P de Araujo, and V.V. da Silva (org.) Cajucultura: Modernas técnicas de produção. Embrapa-CNPAT, Fortaleza.

Ghosh, SN.1995. Studies on effect of watering during flowering and fruiting on yield of cashew. *The Cashew*. 9(3): 5-8.

Ghosh, S.N. 1988. Effect of nitrogen, phosphorus and potassium on flowering duration, yield and shelling percentage of cashew (*Anacardium occidentale L.*). *Ind. Cashew J.*, 19: 19-23.

Ghosh, S.N. 1989. Effect of nitrogen, phosphorus and potassium on flowering duration, yield and shelling percentage of cashew (*Anacardium occidentale* L.) *Ind. Cashew J.*, 19: 19-23.

Ghosh, S.N. 1991. Studies on screening of cashew types for rootstock in softwood grafting. *The Cashew*. 5: 3-4.

Ghosh, S. N.1999. Effect of rain water harvesting on nut yield of cashew. *J. Plantation Crops*. 27(3): 225–229.

Ghosh, S.N. and Bose, T.K. 1986. Nutritional requirement of cashew (*Anacardium occidentale L.*) in lateritic tract of West Bengal. *Ind. Cashew J.*, 18: 11-16.

Ghosh, S.N., Chattopadhyay, N. and Hore, J.K. 2000. Effect of plant spacing on yield of cashew. Recent Advances in Plantation Crops Research 13: 157-159 (Proceedings of PLACROSYM- XIII, 1998).

Gorakhsingh, Hubballi V.N. and Jnanadevan, R. 2016. Good Agricultural Practices in Cashew. Technical Bulletin. Published by: Director, DCCD, Kochi-11.

Green, S., McNaughton, K., Wunschem, J.N. and Clothier, B. 2003. Modelling light interception and transpiration of apple tree canopies. *Agron. J.*, 95: 1380-87.

Grundon, N.J. 1999. Cashew nuts in North Queensland respond to phosphorus and sulphur fertilizers. *Better Crops International*. 13: 22-24.

Gupta, C.R. 1999. Intercropping in cashew orchard under rainfed condition- a model for Bastar plateau zone of Madhya Pradesh. *The Cashew*. 13: 18-22.

Guruprasad, T.R., Hanumanthappa, M., Sudhir, K.V. and Dhananjaya, B. 2009. Technologies for improvement of senile cashew gardens. *The Cashew and Cocoa J.*, 1: 18-22.

Haag, H.P., Sarruge, J.R., de Oliveira, G.D., Scoton, L.C. and Dechen, A.R. 1975. Nutrição mineral do cajueiro (*Anacardium occidentale* L.). Absorção de nutrientes - nota prévia. Anais da E.S.A. "Luiz de Queiroz" 32: 197-204.

Hammed, L. A. and Adeyemi, E.A. 2008. Germination and seedling performance of cashew (*Anacardium occidentale,* L.) as affected by nut-sowing orientations and cotyledon removal. *Nigerian Journal of Horticultural Science (NJHS)*. 10: 59-64.

Hanamashetty, S.I., Hedge, N.K. and Basavaraj, N.2002. Performance of cashew varieties under high range zone of Karnataka, Plantation Crops Research and Development in the New Millennium (eds. Rethinam, P., Khan, H.H., Reddy, V.M., Mandal, P.K. and Suresh, K.). Coconut Development Board, Kochi, pp. 351-353.

Harris, C.V., Pandian, I.R.S. and Thangavelu, S. 1994. Pre-treatment of cashew seeds to improve germination. *South Indian Horticulture*. 42(2): 121-122.

Heading, F. (1992). Cashew establishment and management techniques on Kununurra clay. *Working Papers of the Fifth Annual Cashew Research and Development Workshop, May 18-19, 1992, Kununurra, Western Australia*. 25-7.

Jayalekshmy, V.G. and John, P.S. 2005. Pollination studies in cashew. *Indian Cashew* 19(1): 20-27.

Jayaprabhavathi, S., Keisar Lourdusamy, D., Vincent, S. and Aneesarani, M. S. 2016. Efficacy of curative treatments using insecticides against cashew stem and root borer, *Plocaederus ferrugenius* L. (Coleoptera: Cerambycidae). Serials Publications, ISSN: 0254-8755. *International Journal of Tropical Agriculture,* 34(1): January-March: 13-16.

Jeeva, S., Aneesarani, M.S., Ambethgar, V. and Vaidyanathan, R. 2015. Cashew Germplasm of Tamil Nadu: Characterization and Cataloguing. *Acta Hortic*. 1080: 151-154.

Jnanadevan, R. 2016. Trend in Indian Cashew Trade Flow. *The Cashew and Cocoa Journal*. 5(3): 26-29.

JoseMathew and Mini. C. 2008. Cashew apple. Economic utilization through value addition. In: Recent initiatives in Horticulture (Eds. K.L.Chadha, A.K.Singh and V.B.Patel. Horticultural Society of India, New Delhi. Pp. 612-623.

Kadam, S.G., Nawale, R.N and Nadkarni, H.R. 1996. In-situ softwood grafting in cashew: effect of size and maturity of scion sticks, var. Vengurla-4. *The Cashew*. 10(1): 18-21.

Karihaloo, J.L. and Archak, S. 2000. Molecular markers as aid to agro biodiversity management: some Indian case studies. Extended Summary (2) Voluntary papers. International Conference on Managing Natural Resources for sustainable agricultural Production in 21st Century, 14-18, Feb. 2000, New Delhi.

Keisarlourdusamy, D., Aneesarani, M.S. and Jaya Prabhavathi, S. 2016. Experiments on intercropping young cashew orchards with direct sown vegetables (In Tamil) In:Scientific Agriculture in Tamil-2 Eds. K.Ramasamy, P.Pandiarajan, M.Jawaharlal, S.Sekar and M.Manimekalai. ISBN: 81-902877-3-7 Publ. by AC&RI, Trichirappalli pp. 963-965.

Keisarlourdusamy, D., Aneesarani, M.S. and Jaya Prabhavathi, S. 2016. Experiment on organic management of cashew In: Payir

Melanmaiyiyal (Tamil): ISBN:978-93-84234-55-3 Eds.: P. Thukaiyannan, N.Kiruthiga, D.LeninRaja and S.Senbagavalli and S.Elamathi) p. 331-336.

Kernot, I. (Coord.). 1998. Cashew: Information kit. Queensland: Department of Primary Industries. (Agrilink series QAL 9806).

Keshavachandran, R. and Riji, V.S. 2005. Rapid multiplication of cashew through *in-vitro* methods. Proc. National Symposium on Biotechnological Interventions for Improvement of Horticultural Crops: Issues and Strategies. 10-12 January, 2005, Kerala Agricultural University, Thrissur, pp. 204.

Khan, M.M., Kumar, D.P. and Hiremath, I.C. 1987. Pruning studies in cashew: *The Cashew*. 1: 11-13.

Khan, M.M., Hegde, M., Hiremath, I.G. and Mallika, B. 1985. Rejuvenation of old cashew trees by topworking. *Ind. Cashew J.*, 17: 9-25.

Khan, M.M., Hiremath, I.G. and Kumar, D.P. 1988. Topworking in cashew–A rapid method of producing scion material. *Cashew Bull.*, 25: 11-13.

Kumar, D. P., Subbarayappa, A., Hiremath, I. G., Khan, M. M. and Sadashiviah .1989. Use of coconut coir-pith - A biowaste as soil mulch in cashew plantations. *The Cashew*. 3: 23-24.

Kumar, D.P. 1990. The growth and yield performance of top-worked cashew trees. *Cashew Bull.*, 27: 7-10.

Kumar, D.P., Hegde, M. and Guruprasad, T.R. 1998. Fertigation for higher nut production in cashew (*Anacardium occidentale* L.), *Cashew Bull.*, 35: 2-4.

Kumar, D.P., Khan, M.M. and Melanta, K.R. 1994. Effect of growth regulators on sex expression, fruit set, fruit retention and yield of cashew. *Proc. of the XI Symposium on Plantation Crops,* 30 November -3 December 1994, Calicut, Kerala, India, pp. 610-627.

Kumar, D.P., Subbarayappa, A., Hiremath, I.G., Khan, M.M. and Sadashivaiah. 1989. Use of coconut coir-pith: a biowaste as soil mulch in cashew plantations. *The Cashew:* 3: 23-24.

Kumar, N., Ponnuswami, V., Jeeva, S., Ravindran,C. and Kalaivanan, D. 2012. Cashew Industry in India–An Overview. *Chronica Horticulturae.* 52(1): 23-29.

Lakshmipathy, R. 2000. Symbiotic response of cashew root stocks to different VAM fungi. *The Cashew*. 14: 20-24.

Latha, A and Abdul Salam, S. 2003. Response of cashew seedling to soil moisture stress. *The cashew*. 17(3): 17-21.

Latha, A. and Salam, M.A. 2001. Cashew varieties tolerant to K-deficient soils. *J. Plant. Crops*. 29: 16-21.

Latis, T., and Chibiliti, G. 1988. Foliar diagnosis of nutrient deficiencies in cashew: A study conducted in the Western Province of Zambia. *Revista di Agricoltura Subtropicale e Tropicale*. 82: 677-689.

Lima, R. de L.S. de., Fernandes, V.L.B. and Oliveira, V.H. de. 2001. Growth of seedling dwarf-precocious cashew influenced by organic and mineral fertilizations. *Revista Brasileira De Fruticultura*. 23: 391-395.

Mahapatra, G. and Bhujan.1974. Land selection for cashew plantation- A survey report. *Cashew Bull*. 11(8): 8-15.

Mahesha, U.S., Hanamashetti, S.I. and Hedge, N.K. 2005. Performance of cashew (*Anacardium occidentale* L.) varieties under hill zone of Karnataka. *The Cashew*. 19(3): 5-7.

Mane, M.S., Mahadkar, U.V., Ayare, B.L. and Thorat, T.N. 2009. Performance of mechanical soil conservation measures in cashew plantation grown on steep slopes of Konkan. *Ind. J. Soil Conservation,* 37: 181-184.

Manjunatha, D. 2001. Effect of nutrients (NPK) supplied through irrigation water on growth of rootstocks and grafts of cashew. *The Cashew*. 15: 13-18.

Mantell, S.H., Boggetti, B., Bessa, A.M.S., Lemos, E.P., Abdelhadi, A. and Mneney, E.E. 1997.Micropropagation and micrografting methods suitable for international transfers of cashew. **In:** *Procs. international Cashew and Coconut Conference,* Dar es Salam, Tanzania, February 17-21, p. 24.

Maranan, F.S. and Mendioro, M.S. 2008. Isozyme analysis of Philippine cashew (*Anacardium occidentale* L.) germplasm. *The Philippine Agri. Scient.,* 91: 329-333.

Mengel, K., and Kirkby, E.A. 1978. Principles of plant nutrition. International Potash Institute (IPI), Horgen, Switzerland.

Menon, M.A., and Sulladmath, U.V. 1982. Mineral nutrition of cashew. Indian Cashew Journal: 14: 7-13.

Mika, A.1986. Physiologial responses of fruit trees to pruning. *Horticultural Reviews* (Ed. Jules Jenic K) Vol. 8: 339-378.

Millington, A. J. (1992). Cashew research: Overview Western Australia. *Working Papers of the Fifth Annual Cashew Research and Development Workshop, May 18-19, 1992, Kununurra, Western Australia*. 1-6.

Mini, C., John, P.S. and Sushama, P.K. 2005. Vermicompost preparation from organic wastes of cashew garden. *The Cashew*. 19(4): 15-19.

Mishra, J.N., Paul, J.C. and Pradhan, P.C. 2008. Response of cashew to drip irrigation and mulching in coastal Orissa. *J. Soil and Water Conservation*, 7: 36-40.

Mneney, E.E. 2001. In-vitro micrografting of cashew. *Plant Cell, Tissue and organ Culture*, 66(1): 49-58.

Moffat, A.S. 2000. Can genetically modified crops go 'Greener'? *Science*, 290: 253.

Mohan, E. 1983. Studies on the effect of pruning and growth regulators on vegetative growth and flowering in cashew (*Anacardium occidentale L*) M.Sc (Agri) thesis.

Mohan, E. and Rao, M.M. 1995. Effect of growth regulators and pruning on the growth and yield of cashew. *Environ. Eco.*, 13: 675-679.

Mohan, E. and Singh, R. 1988. Effect of time and severity of pruning in cashew. *J. Plantation Crops* (Suppl.) 18: 7-10.

Mohapatra, A.R., Vijayakumar, K. and Bhat, N.T. 1973. A study on nutrient removal by the cashew tree. *Indian Cashew J.*, 9: 19-20.

Mohapatra, R.N. and Jena, B.C. 2008. Effect of prophylactic measures in management of cashew stem and root borer (*Plocaederus ferrugineus* L.) *J. Plantation Crops*. 36(2): 140-141.

Murali, K. 2012. Effect of tertiary shoot pruning and foliar spray on yield and quality of cashew (*Anacardium occidentale L.*) under high density planting system, M.Sc Thesis, Department of spice and plantation crops, HC&RI, TNAU, Coimbatore.

Murali, K., Prasannakumar, P. and Aneesarani, M.S. 2015. Effect of tertiary shoot pruning and foliar spray of nutrients on flowering and yield of cashew (*Anacardium occidentale L.*) under high density planting system, In: *The Bioscan*. 10(1): 411-415.

Nagaraj, K.V. 1998. Quality and processing aspects of cashew. Harvest and Post- harvest Technology of Plantation Crops,(eds. Bosco, S.J.D., Sairam, C.V., Muralidharan, K. and Amarnath, C.H. CPCRI, Kasargoid, pp. 111-113.

Nalini, P.V.1997. Productivity in relation to branching pattern and pruning in cashew (*Anacardium occidentale L.*), Ph. D Thesis, Kerala Agricultural University, Thrissur, 147p.

Nambiar, M.C. and Pillai, P.K.T. 1985. Cashew. *In* Fruits of India: tropical and subtropical. *Edited by* T.K.Bose. Naya Prokash, Calcutta, India. pp. 409–438.

Nawale, R.N., Sawke, D.P. and Salvi, M.J. 1985. Effect of black polyethylene mulch and supplemental irrigation on fruit retention in cashewnut. *Cashew Causerie*. 7: 8-9.

Nayak, M.G. 1996. Training and pruning practices for cashew. *The Cashew*. 10: 5-9.

Nayak, M.G., Mohana, G.S., Bhat, P.S., Saroj, P.L., Swamy, K.R.M. and Bhat, M.G. 2014. Minimum Descriptors of Cashew Germplasm Accessions: Catalogue-V Published by : Prof. P. L. Saroj, Director, Directorate of Cashew Research, Puttur - 574 202, Dakshina Kannada, Karnataka.

Neto, P.S, Maruta, I., Takaiwa, F., Oono, K., Matsumoto, K. and Gomes, J.A. 1995. Identification of cashew (*Anacardium occidentale* L.) seedlings with RAPD markers. *Acta Hort*. 370: 21-26.

Nivas, S.K. and D'Souza, L. 2004. Somatic variation in tissue cultured derived mature cashew trees. *J. Plantation Crops,* 32: 144-146.

Oblisami, G., Santhanakrishnan, P., Pappaiah, C.M. and Shanmugavelu, K.G. 1985. Effect of *Azotobacter* inoculant and growth regulators on the growth of cashew. *Acta Hort.,* 108: 40-45.

Oblisami, G., Santhanakrishnan, P., Pappaiah, C.M. and Shanmugavelu, K.G. 2011. Effect of *Azotobacter*inoculant and growth regulators on the growth of cashew. *Acta Hort.,* 108: 40-45.

Ohler, J.G. 1979. *Cashew*. Koninklijk Instituut voor de Tropen, Amsterdam, 260p.

Oliveira, V.H. de., Santos, F.J. de S. and Miranda, F.R. 2002. Fertirrigação. p. 89-94. *In*: L. de M. Barros (org.) Caju Produção: Aspectos técnicos. Embrapa Agroindústria Tropical. Fortaleza.

Opoku-Ameyaw, Oppong, K F.K., Amoah, F.M., Osei-Akoto, S. and Swatson, E. 2011. Growth and early yield of cashew intercropped with food crops in northern Ghana. *Journal of Tropical Agriculture*. 49(1-2): 53-57.

Parente, J.I.G., Maciel, R.F.P. and Vale, E.C. 1972. Cajueiro: Aspectos econômicos e agronômicos 2nd ed. IPEANE. Recife, PE.

Peng, R.K., Christian, K. and Gibb, K. 1995. The effect of green ant, *Oecophylla smaragdina* (Hymenoptera: Formicidae) on insect pests of cashew tree. *Bulletin of Entomological Research.* 85: 279-284.

Peter, K.V. 2002. Cashew - Plantation Crops. National Book Trust. p.55-80.Satapathy, B. (1988). Thinning, training and pruning of cashew to increase production and and productivity of cashewnut. *The Cashew.* 2: 6-8.

Pillai, G.B., Dubey, O.P. and Singh, V. 1976. Pests of cashew and their control in India. A review of current status. *J. Plantation Crops.* 4: 37-50.

Pushpalatha, P.B. and Kutty, M.C.N. 2005. Cashew apple processing, an analysis. In: Cashew apple processing for increased income (Eds. Jose Mathew and Abraham, M). Cashew Research Station, Kerala Agricultural University, Madakathara, p. 1-6.

Rajangam, J., Azhakiamanavalan, R.S., Thangaraj,T., Vijayakumar, A. and Muthukrishnan, N. 2001. Status of production and utilization of moringa in southern India. **In:** Seminar on development potential for moringa products, October 29th - November 2nd, 2001, Dar es salaam, Tanzania.

Ramesh, N., Lingaiah, H.B. and Khan, N.M. 1999. Economics of bio-fertilizer application in cashew graft production under eastern dry zone of Karnataka. *Crop Res.,* (HISAR) 17: 336-341.

Rao, E.V.V.B and Swamy, K.R.M. 2006. Cashew.(In) Handbook of Industrial crops, (eds. Chopra, V.L. and Peter, K.V.). Haworth Press, USA, pp. 77-136.

Raviprasad, T.N. Shivarama Bhat, P. and Sundararaju, D. 2009, Integrated pest management approaches to minimize incidence of cashew stem and root borers (*Plocaederus* spp.). *Plantation Crops.* 37: 185-189.

Reddy, A.V., Narasimha Rao, P.V., Andaih, S. and Bubba Rao, I.V. 1981. Cashew NPK nutrition in relation to growth under graded doses of nitrogen fertilization. *Indian Cashew Journal.* 14: 15-21.

Russell, D.C. 1969. Cashew nut processing, by FAO Agricultural Services Bulletin No. 6.

Salam, M.A. 1999. Cashew varieties suitable for high density planting. *Cashew Bull.,* 37: 35.

Salam, M.A. and Joseph, M.P. 2003. Apple and Juice yield potential of cashew varieties. *Journal of Plantation crops.* 31(3): 18-22.

Samiayyan, K., Palanisamy, K.P., Ahamed Shah. H. and Manivannan, K. 1991, Effect of Prophylactic measures against cashew stem and root borer (*Plocaederus ferrugineus*) *The Cashew*. 5(1): 16-17.

Santos, F.J. de S., Miranda, F.R., Oliveira, V.H. de. and. Saunders, L.C.U. 1997. Irrigação localizada: Microirrigação, Fortaleza: Embrapa-CNPAT. (Documentos, 23).

Sapkal, B.B. 2000. Integration of foliar application of urea and insecticidal spray for yield maximization in cashew. *The Cashew*: 14: 41-44.

Sathpathy, B. 1988. Thinning, training and pruning of cashew to increase production and and productivity of cashewnut. *The Cashew*. 2: 6-8.

Schaper, H., Chacko, E.K. and Blaikie, S.J. 1996. Effect of irrigation on leaf gas exchange and yield of cashew in northern Australia. *Australian Journal of Experimental Agriculture*. 36: 861-868.

Senguttuvan, T. 1999, Distribution and intensity of infestation by cashew stem and root borer (*Plocaederus ferrugineus*) L. in Tamil Nadu. *Pest Management in Hort. Ecosystem*. 5(2): 154-156.

Shi-WenGe., Qiao-GuanMing., Wang-YongGang. and Zhu- XinTian. 2005. Experiment of applying nitrogen and phosphate fertilizers for cashew nut. *South-China Fruits*. (1):30 (Research Institute of Forestry, Honghe Harni and Yi Nationality Autonomous Prefecture, Yunnan, China).

Singh, S.K., Syamal, M.M. and Maurya, A.N. 1992. Effect of NAA and Ethrel on vegetative growth, flowering, fruiting and yield of cashew nut. *The Cashew*. 6: 11-12.

Singh, S.N. 1997. Effect of bio-fertilizer (*Azotobacter chroococcum*) in cashew plantation. *Environ. Ecol.*,15: 482-484.

Sivaprasad, P., Sulochana, K.K., George, B. and Salam. M.A. 1992. Growth and phosphorous uptake of cashew as influenced by inoculation with mycorrhizae. *The Cashew*, 6: 16-18.

Sivaraman, K. and Hubballi, N. 2002. Organic Farming in Cashew' in Indian Cashew:

Sridhar, R., Shah, H.A. and Manivannan, K. 1990. The effect of VA mycorrhizae inoculation on certain growth characters of cashew. *Cashew Bull.*, 27: 15-16.

Subbaiah, C.C., Manikandan, P. and Joshi, Y. 1986. Yellow leaf spot of cashew: a case of molybdenum deficiency. *Pl. and Soil*, 94: 35-42.

Sundararaju, D. 2004. Evaluation of certain insecticides against tea mosquito bug on cashew. *The Cashew*. 18(2): 22-25.

Swamy, K.R.M. 1995. Top working of cashew in Goa and Maharashtra States - a case study. *The Cashew*. 9: 12-17.

Swamy, K.R.M., Bhaskara Rao, E.V.V., Nagaraja, B. and Nayak, M.G. 1993. Commercial propagation of cashew varieties. **In:** *Proc. of the Golden Jubliee Symp. on Horticulture Research - Changing Scenario*. Bangalore from 24-28 May 1993. Abstr. 26-31, pp. 448–449.

Swamy, K.R.M., Rao, E.V.V.B., and Bhat, M.G. 1998. Catalogue of minimum descriptors of cashew (*Anacardium occidentale* L.) germplasm accessions-II. National Research Centre for Cashew, Puttur, Karnataka, India.

Thimmappaiah and Shirly Raichal Anil 2000. Micropropagation in mature tree of cashew (*Anacardium occidentale* L.) **In:** Recent Advances in Plantation Crops (Eds: N. Muralidharan and R. Raj Kumar), Allied publishers Ltd., New Delhi. Pp. 45-48.

Thimmappaiah, Shirly, R.A. and Sadhana P.H. 2002. *Invitro* propagation of cashew (*Anacardium occidentale* L) from shoot explants of young trees. *In-vitro Cell. Dev. Biol. Plant*, 38: 152-156.

Thimmappaiah., Melwyn, G.S., Shobha, D. and Shirly, Raichal Anil, 2009. Assessment of cashew species for molecular diversity. *J. Plant. Crops*. 37: 146-151.

Thirumalaraju, G.T., Puttusamy. and Narasimha Reddy, M.N. 2005 Population dynamics of tea mosquito bug (*Helopeltis antonii* Sign) on different age group trees of cashew. *The Cashew*. 19(1): 16-19.

Venkateswarlu, D., Murthy, K.K.V.M., Reddy, L.N.M. and Lakshmi, S.M.M. 2003. Effect of anthracnose disease of the cashew seedlings on the graft in cashew nurseries. *The cashew*. 17(2): 22-24.

Yadukumar, N. 2007. Organic farming in cashew. Proceedings of the National Seminar on Research, Development and Marketing of Cashew, 20-21st November, Goa, pp. 36-46.

Yadukumar, N. and Balsimha, D., 2006, Effect of drip irrigation and fertilizer levels on photosynthesis in cashew. *Indian J. Hortic.*, 63(3): 310-315.

Yadukumar, N. and Mandal, R.C. 1994. Effect of supplementary irrigation on cashewnut yield and water management for Plantation Crops - Problems and Prospects, Centre for Water Resource Development and Management, Calicut, Kerala. pp. 79-84.

Yadukumar, N. and Nandan, S.L. 2005. Recycling organic wastes of cashew plantations by aerobic composting. *J. Plant. Crops.* 33: 99-102.

Yadukumar, N. and Rejani, R. 2006. Temporal and spatial variation of soil moisture content and development of a yield model based on soil moisture content. *J. Plant. Crops.* 34: 364-367.

Yadukumar, N., 2003, Strategies for improved productivity through efficient water management. *The Cashew.* 17(3): 25.

Yadukumar, N., Rao, E.V.V.B. and Mohan, E. 2002. High density planting of cashew. *Trop. Agri.,* 78: 19-28.

Yadukumar, N., Raviprasad, T.N., Nagaraja, K.V., Haldankar, P.M., Godase, S.K., Susanamma, K., Gajendran, G., Mahalingam, T., Lenka, P.C., Mohapatra, R.N. and Bandopadhyay, B. 2003. National Agricultural Technology Project. Final Report on developing integrated production packages for enhancing productivity of cashew. National Research Centre for Cashew, Puttur, D.K., Karnataka. 95pp.

Yadukumar, N., Rejani, R. and Nandan, S.L. 2008. Studies on green manuring in high density cashew orchards. *J. Plant. Crops.* 36: 265-269.

Yadukumar, N., Rejani, R. and Nandan, S.L. 2009. Studies on green manuring in high density cashew orchards. *J. Plant. Crops.* 36: 265-269.

Web references

http://dccd.gov.in/content.aspx?mid=1075&tid=1

http://www.nutrition-and-you.com/cashew_nut.html

https://hort.purdue.edu/newcrop/morton/cashew_apple.html

http://www.vegetafruit.com/2015/12/health-benefits-of-cashew-fruit.html

NABARD website

❑❑❑